The Color Of Oil

The History, the Money and the Politics of the World's Biggest Business

Michael Economides
Ronald Oligney

Original artwork by Armando Izquierdo

Round Oak Publishing Company
Katy, Texas

ROUND OAK PUBLISHING COMPANY, INC.
1811 Breezy Bend
Katy, Texas 77494

Printed in the United States of America

10 9 8 7 6 5 4 3 2 1

Library of Congress Cataloging-in-Publication
Economides, Michael J.
The color of oil : the history, the money, and the politics
of the world's biggest business /
Michael Economides, Ronald Oligney ; with original
artwork by Armando Izquierdo. -- 1st ed.
p. cm.
Includes bibliographical references and index.
LCCN: 99-80030
ISBN: 0-9677248-0-5
1. Petroleum industry and trade.
2. Petroleum industry and trade--History.
I. Oligney, Ronald.
II. Izquierdo, Armando.
III. Title.

HD9560.5.E23 2000 338.2'728
QBI99-901816

For our colleagues

Contents

Preface

This is a story of a human enterprise that has shaped and will continue to shape civilization. Energy and petroleum, *oil* in particular, have caused world changes and generated immense wealth for both producers and users.

It is also a story of people and an industry of many hues—from the green of money; to the black of the physics; to the red, white and blue of America's pervasive influence; to the red of war. Other colors—more complex and not as emphatic, but by no means insignificant—speak of the Byzantine complexity of the business, its profound international character, its hostile-to-chummy relationship with government, and the frequently unfathomable hostility of its enemies.

The industry and, especially, the commodity that it shepherds are here to stay and thrive for centuries. It is literally the royalty of the present and most certainly the future of humanity. Its final color is imperial—purple.

Michael Economides and Ronald Oligney
Houston
January 2000

Foreword

In February 1999, a front-page article in the Sunday op-ed section of the Houston Chronicle talked about the prospects of $30-per-barrel oil within a year—quite a revelation, considering that oil prices at that time were languishing at around $11 per barrel, the lowest price in recent memory.

I was impressed by the reasoning of the authors, invoking the physics and economics of oil production, and even more so, by their optimism about the future of the oil industry.

Although I had never met Michael Economides and Ron Oligney, I called both of them at their homes and gradually a relationship emerged. I had several inspiring conversations with them, and when they asked me to write the foreword to their book, "The Color Of Oil," I accepted with enthusiasm. (Incidentally, as I write this in mid-November, 1999, the price of oil has climbed to more than $27 per barrel.)

I am an oilman by choice. I think that the business, both today and particularly in the long run, is still the El Dorado of all investments. I am also an oilman by conviction. I believe that providing energy to society is a most rewarding and humanly worthy enterprise.

I come from a family of oilmen. My father, Marvin Davis, made his fortune first in the oil fields of the Mid-continent, and eventually in the Midwest, the U.S. Gulf Coast and, finally, all over.

Our story is purely American, from very humble beginnings to success—made all the more striking because of its intimate connection with oil.

My grandfather, Jack Davis, an Englishman, jumped ship in New York in the 1920s, and after working at a number of things, he started a women's dress business. He would make dresses and sell them for $5 each to the nickel-and-dime stores in New York. It was a tough, competitive business, but it paid enough to feed the family.

In 1939, my grandfather met a guy who had an opportunity to drill some oil wells in Illinois and Indiana, and although my grandfather did not know much about oil, the idea appealed to him. In the early 1940s, my father, who had started working for the family dress business, went to the oil fields to check out the investment. It didn't take long for him to catch the bug.

My father got more and more involved in the business, and around 1950, he was introduced to the management side of operations, which he eventually took over. In 1960, he moved to Denver and set up operations there. Three years later, I was born.

During those heady years, my father got advice from legendary wildcatters like Bunker Hunt and H.L. Hunt. I can't tell exactly what it was that they taught him, but they certainly founded his philosophy: *Drill as many wells as you can. It is a gamble, and if you want to be successful, you must take a lot of chances and drill a lot of wells.*

To get as many wells as he could, he perfected the idea of bringing in other partners; instead of drilling one expensive well for $10 million, why not drill 10 wells and limit your risk to $1 million per well?

He developed what is known as "one third for one fourth," which eventually became the industry standard. Three partners would each invest one-third of the cost of drilling in exchange for a one-fourth interest in the profit; the company retained the remaining quarter-share of profits.

In this way, my father built his wildcat empire. By 1980, Davis Oil was one of the top independent operators in the country, having drilled more than 11,000 wells. Between 1976 and 1980, only Exxon, Amoco and Shell drilled more. The company explored in almost all petroleum regions and expanded operations from Denver to Houston to Midland to New Orleans and Tulsa. The number of employees grew to 500.

From as early as I can remember, my father was this jovial big bear of a man. He is the most incurably optimistic man I have ever known, and I think that is what made him so successful. In a business as risky as the oil business, optimism is necessary. This is why, to this day, he is able to do business with any company and anybody.

Optimism is a characteristic of all successful wildcatters. It is essential for the main player and infectious to all who work with him. One must have optimism because, with a lot of discovery, there are also a lot of dry holes. That's the nature of the business.

Of course, how my father maintained optimism after seven or eight dry holes in a new trend was mystifying to me. This is really what eventually lured me into the business.

When I was a kid, my father would take me with him to work on Saturdays. He had very loyal employees, but he gave them great incentives. My father is the kind of a guy who always picked up the phone; he would never take a message. He would simply talk to people. His door was always open. If a geologist had an idea, my father would simply ask, "How much is it? Let's go get it."

So, I watched my father build his fortune.

With success came investment opportunities in other businesses. He created a lot of relationships with people, and these relationships offered him great opportunities in other businesses. He built a strong real estate business in Denver that grew to become a nationwide enterprise, constructing Class A office buildings in many metropolitan areas. Whenever he saw a good opportunity, he jumped on it.

One day, somebody asked my father if he was interested in the movie business. Twentieth Century Fox was up for sale. Now, Dad loved movies. In our new house in Denver, he had built a small movie theater. He also loved California. So, he became one of the first owners of a major privately held studio. After the oil business hit its low point in the mid-1980s, he packed the family and we moved to Los Angeles.

There are many similarities between the oil business and the movie business. Oil starts with ideas, movies start with ideas; oil has wildcats, the movies have wildcats, too.

For instance, in the Wilcox formation in Texas, there are different structural trends. Some still haven't been drilled. A star geologist may suggest drilling in one, but no matter how much science and technology you have, no matter how much experience you have, you can never be sure of a successful strike.

The same is true of Hollywood. There are also trends. Everybody loves Walt Disney and the cartoons; everybody likes good cop movies or big-budget action adventures. Big-name stars are also supposed to play a role. And yet, movies can also be boom-and-bust.

Both businesses require a strong stomach. Does anybody think that he can go to Hollywood and invest in only one movie and be a success? The same things apply in the oil business.

My brother, who stayed in the movie business after my father sold the studio, has gone on to produce several successful movies. Like a good oil well, those movies began with great ideas and had all the right ingredients, trends and big-name stars. However, in the oil business and in the business of successful producers/wildcatters in Hollywood, sometimes all those ingredients still don't work, and you get a dry hole.

For myself, while growing up, I had ideas of doing all sorts of other things. Then, I thought about what a great business this is. The rewards are tremendous. The lure of "the green" is enormous. The people I've met have been very interesting—Occidental's Armand Hammer with his great worldly scope, the flamboyant Ray Plank, head of Apache, and the venerable and wise Oscar Wyatt of Coastal. These men built great fortunes and great companies.

So now, I am in the oil business. Over the years, Davis Oil, which became Davis Petroleum, went up and down with the oil business. In the early 1980s, my father sold a majority of his petroleum business at a great price. Shortly afterward, the oil business went south. The staff was downsized, but we never lost our optimism.

I always thought that the price of oil, now used as a commodity, could go down to $10 per barrel for a while. We are set up to do business at that price. However, we can make a lot more money at $30 per barrel. The same is true for gas. We can survive at $1.90 or even $1.50 per thousand standard cubic feet. Our profits, of course, are much better at $3 per thousand standard cubic feet. One can make it in the business with a lot fewer, but better people. Recently, we had some very nice discoveries with a handful of people.

A great deal is said today about the information revolution. But what underlies the whole thing? Basic necessities—food, water and, for sure, energy. I believe that all revolutions, from the industrial revolution to today's information revolution, are founded on energy. Nobody can turn a computer on without it.

The spirit is that the business will always be here. The world runs on petroleum and whoever controls some of it will do well. I have that mentality.

"The Color Of Oil" is a compelling read for anyone interested in understanding the industry, its past, present and future. It will become a classic.

Gregg Davis
Houston
November 1999

Part I
Green

The money, wealth and economics of oil

A classic scene from Shoestring Alley at Sour Lake, Texas, gives a striking impression of the oil industry circa the early 1900s. *(Courtesy: American Petroleum Institute)*

The temperature in the desert is frying, close to 50 degrees Celsius (122 F) in the shade … it never goes higher than 50 degrees they tell me, because the government has decreed that if it does, then work must stop. There is, of course, no way that its national oil company can stop drilling for oil. That's why the temperature can never climb above 50 C.

Inside the driller's "office," the air-conditioning unit, itself probably never reconditioned, does a barely acceptable job, making occasional gurgling noises; it could die at any moment.

Outside, the tool pusher and his crew, manning the drilling rig floor and the rig surroundings, go about their business—macho, perverse in their not complaining, sweating and unwashed for days. They look more like hippies, for certain not the San Francisco variety, all with their cowboy boots and the slow drawls of Midland or Henderson, Texas, and Muskogee, Oklahoma.

The perimeter of the rig, with the fine sand of the Arab desert showing under the heavy equipment and supplies, is ringed by a brown goo. As I stand there wondering for a couple of minutes, my curiosity is quickly replaced by a young engineer's disgust. And then, it is both funny and understandable. You mature in the oil field if you can recognize the unmistakable trace of tobacco spit, and it no longer disgusts you.

Change temperature and color but not so much the scenery, flat forever with only man-made relief. Today, everything beyond 50 feet is invisible. The whiteout blends sky and earth in a deafening whiteness. It is the North Slope of Alaska. The temperature outside is something ridiculous, and I no longer care to distinguish between wind-chill index and real temperature, which is perhaps 100 F—below zero.

And here, where the opposite extremes of earth are found, here you find the commonality of the drilling rig and its crew, the buddies of those in Arabia. Some have been there themselves. I hear the drawl, and I see the cowboy boots and the tobacco spit again. Everybody in Alaska knows that you don't eat yellow snow, but for sure, you don't want to touch brown snow.

The men of these rigs, sons of beaten-up farmers and failed junker-car mechanics, found their fortune in the oil business, some, in good years, making money that they could never have dreamt of in the Texas prairie towns. The money they made for the owners of the oil wells was just beyond belief. – M.E.

✧

The color of oil is green, and even if money throughout the world has all the colors of the rainbow and then some, it is the greenback, both literally and figuratively, that has defined the value of oil.

Money—lots of it—has always been associated with oil, from the public's imagination to the perpetual *nouveau riche* mentality of Texas oil barons to the ostentatious spending in the Arab Gulf countries. Until Bill Gates took over because of the crazy run on "technology" stocks, for years, the richest man in the world has been the Sultan of Brunei, who was always referred to as "sitting on top of a sea of oil."

This image has not always been good.

Striking it rich through a generous dose of good luck and not so much hard work has created envy, jealousy, and at times, righteous indignation. The size of the industry and the vital importance of its product and services emote both admiration and fear.

First, there was John D. Rockefeller, to this day the "I, monopolist" who helped make antitrust a fixture of the American political scene. Standard Oil and its breakup in 1911 defined forever the inherently contradictory, very capriciously applied and economically inconsistent principles that defied capitalism (or "free enterprise," depending on one's point of view) and, at the same time, threw bones to social engineering.

President Theodore Roosevelt was not a socialist by any means, but his campaign against Standard Oil and Rockefeller himself strengthened the government's role in public life and the economy. Government's role as benign overseer and protector of the common man against evil and greedy business lasted for more than half a century.

The Standard Oil breakup proved to be a quick boost for the industry. Not only did the breakup cause Standard Oil to relinquish control over its engineers and executives; it also gave people, both in the United States and elsewhere, the idea that there was an opportunity in this business after all.

The uniquely American notion that landowners also owned the subsurface mineral rights found its highest expression in oil, and in Texas. Texas was largely spared by Standard Oil, which concentrated in the East and the Midwest and had only sparse production in the late 19th century. Then, around New Year's Day, 1901, the Spindletop discovery well near Beaumont, Texas showered a gusher of oil at a rate of 100,000 barrels per day, an amount that very few countries produce today.

No gold rush ever matched the influx of fortune hunters, land speculators and simple job seekers that the oil discovery in Texas attracted. The accumulations of the resource—even if the geology could rationalize it—were finicky, lopsided, and, depending on your land, either better than a pot of gold or abjectly unfair.

Nouveau Riche

Those who got rich became filthy rich. The century of oil in Texas created the *nouveau riche*—something that the world had never seen before, or certainly not at that scale. Immigrants to the United States came from neither the intellectual nor the moneyed aristocracies of the world. To this day in Europe, the rich are born rich, and the chances of building a fortune from scratch are nearly as remote as winning the lottery. Suddenly, oil made ranchers (with practically no education and no social status) richer than the richest European aristocrats and the few upper-class Easterners of the United States.

This image has stuck in the popular fancy. Hollywood obliged and also struck it rich, with productions from the "Beverly Hillbillies" to the scheming, ruthless, country-simpleton billionaires of "Dallas." Even the notorious London traffic would come to a noticeable lull while "Dallas" was showing.

Parallel to the United States, where exploitation and consumption largely matched each other until the late 1960s, the French, Dutch and British all sent prospectors to find oil, first in their colonies and then everywhere else. And find oil, they did—sometimes in great abundance. Along with large American companies, most of which were offsprings of Standard Oil, they made oil production, refining and transportation—from its sources to the main consumers in the United States and Europe—the largest business in the world, and by far, the most international. This is the world of "Big Oil."

Indonesia, Imperial Russia, Persia (now Iran and Iraq), Nigeria, French Africa (primarily Algeria), Venezuela and Arabia (eventually fashioned into several countries) became the world's biggest oil producers. With relatively little consumption of their own, these nations emerged as the petroleum exporting countries, ultimately formalizing into OPEC.

Decolonization in the 1950s brought national identity in the 1960s, and shortly thereafter, nationalization of the oil industries.

Money came pouring in, and after a fluke of an embargo following the Arab-Israeli War of October 1973, money *really* came pouring in. The oil price that gradually increased from $2 to $3 per barrel between 1970 and 1973, climbed to $5.40 per barrel on October 16, 1973. By November, rumors of Nigerian oil selling for $16 per barrel were affirmed by a stunning mid-December auction of Iranian crude, bringing more than $17 per barrel.

Never before and never since has the world experienced such a massive reorientation of wealth as that which took place over the next decade. Seemingly minor political events, even internal-to-the-producers conflicts, caused prices to spike to more than $40 per barrel. Major oil companies actually published forecasts of $100 per barrel in the 1980s, and throughout the consuming world, panic ensued. Spasmodic movements toward alternative sources of energy and visions of returning to the woods became serious fodder for talk shows, books, and even official U.S. government documents and policy statements.

The term *energy crisis* became part of the industrialized world's lexicon, and both the term and its alluded meaning and implications persisted for almost a decade.

The oil producers, now flush with money, became voracious consumers of everything imaginable and even unimaginable.

Some economists tried to separate these countries into "absorbers" and "nonabsorbers" according to their population sizes. Presumably, "absorbers" could actually spend the money; "nonabsorbers" might not. In truth, all of them spent it all and then borrowed some more, leveraging against future earnings. Banks fell over each other to provide loans.

To the euphoria of the locals and the bemusement-to-hostility of all others, including the gleeful suppliers of the most outrageous toys and playthings, few people in the producing countries made the distinction that absorption did not really mean only spending. Even fewer gave thought to reinvestment, infrastructure, and training of people, or to creating institutions of social, economic and political maturity.

These cultures, with little background in those things that characterized the wealth of nations for centuries, suddenly found themselves with per capita incomes comparable to or larger than those of the most established countries.

Alice in Wonderland may have seen more comprehensible situations than things happening in some of the oil producing states during the oil boom. For a while, one of the most lucrative jobs in Caracas was that of expediting imported goods from the hulking cargo ships that were anchored not just in the harbor, but all along the northern coast of Venezuela. Middle-class Caraceños would not think twice about packing the extended family for a weekend in Florida.

More serious was the influx of country folks, previously eking a living from agriculture and other basic and primitive (but at least productive) pursuits, into the big cities. They were invited by populist politicians who promised a share in the country's newly found riches. Of course, what they generally found were welfare crumbs. Any chance for the countries to evolve gradually was destroyed in the process. When the oil wealth dried up in 1983, the out-of-control shanty towns surrounding Lagos, Jakarta and Caracas became permanent fixtures of squalor, replete with social, economic and civic unrest. These are places where even today the police dare not enter.

Corruption thrived through the days of abundance and was even more pervasive in the lean days following. It still thrives today. Oil producing countries generally comprise the top three on everybody's list of corrupt countries in the world. This should not be surprising in nations that have no democratic institutions or economic infrastructure, and are geographically and culturally far from the egalitarian notions of the United States and Europe that emerged from the industrial revolution.

Beyond the thievery practiced by the governing elites and those aspiring to be elites, corruption often became blended with gross mismanagement. Corruption distorted just about everything: words and terms lost their meanings; government and social institutions were just titles with no relation to functionality; armies and police were instruments of the power elite; and banks became the looting grounds of all.

Nowhere, though, did oil wealth have the impact that it brought on the Arab Peninsula, and nowhere else did money become so synonymous with national identity. Tribal fiefdoms, quickly renamed as countries in vast deserts with indescribable national borders, suddenly found themselves with per capita incomes at the level of Europe's—with a couple of them surpassing the per capita income of the United States.

Of course, calculating per capita income was really fruitless, merely an exercise in arithmetic, because no distinction was made between state coffers and the personal wealth of the tribal leadership. After splurging on numerous palaces trimmed in solid gold bathroom fixtures, the jet-setting, playboy leaders provided their denizens with reckless abandon. Desert nomads became the richest welfare recipients in the world. If one needed anything at all, he only had to ask the ruler personally, in tribal audience, and his wish generally would be granted.

The legendary stories, fantasy and reality blended together, no one ever bothered to check: more Mercedes automobiles per capita than anywhere in the world, tens of thousands of dollars in giveaways to each person per year, and scores of brand-new apartments built for the locals whether they asked or not. Many a new high-rise, built at a premium by foreign companies, remained uninhabited or was used for storage. Desert people are more comfortable under their tents. Stories of Arab princes dropping millions on the gambling tables of Monte Carlo with a couple of blondes on each side were so often repeated that they are bound to be true.

Marshaled-to-spend-the-money did not mean that the locals managed it, and clearly did not imply any work. How many native Kuwaitis or citizens of the United Arab Emirates really exist? What is the true population of Saudi Arabia?

The answers are closely held secrets, not to be discussed. There was a time when asking the question might be construed as subversive. A whole rainbow of nationalities has been recruited to do the work, from European and American expatriates who manned most of the professions in the beginning, to Egyptians and Palestinians in their diaspora, to a host of servile nationalities from the Indian subcontinent and the Philippines.

With no rights and often no protection, with passports confiscated and kept for the duration of the "contract," the imported workers often endured unspeakable indignities. Yet, the alternative incomes in their homelands were a small fraction of what they could earn in their adopted workplaces.

It was not just the immediate ruling families who benefited from the money rush. In many countries, extended families and tribal noblemen made money, hand over fist. One of the most common ways was to facilitate a foreign company's entry into the country. Formally, it was not just foreign company XYZ operating there, even if the logo and the people and the equipment were unmistakably theirs. It was local Al-ABC Enterprises arranging the work for a convenient 15 to 20 percent commission.

Road construction in Saudi Arabia during the oil boom of the 1970s was so constant that the joke was inevitable: "Saudi Arabia built 100,000 kilometers of roads last year. The Kingdom now has a total of 20,000 kilometers of paved roads."[1]

Spreading Out

Worldwide, the huge oil-price increases of the 1970s and early 1980s, and the new technologies of the late 1980s and early 1990s (such as 3-D and 4-D seismic measurements and horizontal drilling) resulted in a resurgence of exploration and production. This resurgence bolstered activities on the North Slope of Alaska and the North Sea, and led to quite a lot of activity in the Gulf of Mexico. Many countries not previously associated with oil production—for example, Papua New Guinea or Barbados—also rushed to find and exploit their oil resources.

In the consuming nations, the "oil crisis" panic was gradually replaced by a realization that oil was not running out and that prices in real dollars were coming down. Led by the major oil companies and a whole bunch of newcomers, retailing of abundant petroleum products became the most visible manifestation of oil. The transportation sector—trains, planes, ships and automobiles—was and is, by far, the biggest user of oil. Some people don't realize that gasoline was selling in nominal dollars for more in 1979 than in 1999, while the price of cars increased by a factor of six during that time.

When the history of the last part of this century is written, the world may finally understand that few political and economic events have held such diverse and unexpected results as the "energy policies" of the Jimmy Carter and Ronald Reagan administrations.

After the 1976 energy crisis, new notions entered the public discourse and national and international ideologies; all of these confused efficiency (the fading away of huge gas-guzzling automobiles) with oil and gas conservation. Along with the clearly ridiculous but romantic notions of wood-burning, or the even more ridiculous and atrociously expensive solar and wind energies (never reaching 1 percent of the energy mix), a certain malaise settled over the United States and the developed world during the Carter administration.

The Reagan administration contrasted this gloomy picture and decreed low oil prices, primarily through deregulation and open markets, as its real and shadow strategy. Low oil prices became the leading edge of a double-edged sword that brought the rapid demise of the Soviet Union and the communist bloc, whose only credible export and source of "hard" currency was oil. The effect became devastating when combined with Reagan's wholesale and expensive-to-match arms buildup, the second edge of the sword.

This did not come without an early cost. Many banks and other financial institutions became overextended both in the United States and especially in South America because of crazy projections of future oil revenues. When oil prices collapsed, the banks did too. The resulting "banking crisis" of the 1980s panicked many, but in the long run, it proved highly beneficial both for the economy as a whole and for the banking sector.

Possibly more important to President Bill Clinton and the average American at the start of 2000 was the fact that low oil prices in the 1980s also fueled an unprecedented economic growth that continued unabated more than a decade later.

Energy Consumption and the Wealth of Nations

Ever since Adam Smith, political economists have tried to relate national wealth to some national propensity or characteristic. Today, the energy wealth and poverty of nations has replaced industrialization as the defining factor. A robust economy is marked by very large per capita energy consumption (Figure 1). Moreover, demand for energy does not result from wealth, but instead, promotes and generates wealth.

Oil and gas account for 60 percent of the world's energy needs, a market share that has remained largely constant in the past 25 years and one that will probably increase in the next century. Including coal's 30 percent share, the hydrocarbon mix comprises more than 90 percent.

Low oil prices that continued through the 1990s are perhaps the single most important reason that the diverse and developed economies of the United States and Europe have flourished. Low oil prices and abundant energy have led to one of the clearest notions of national security and the establishment of the United States as the single, undisputed world power.

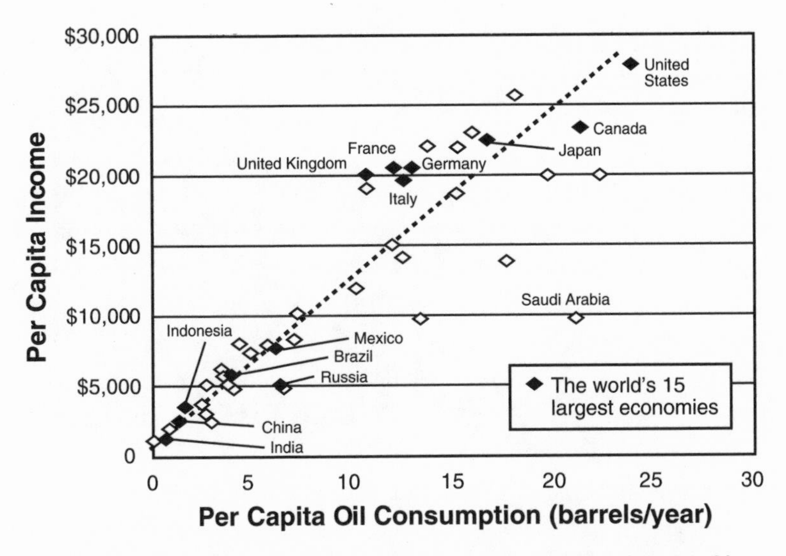

Fig. 1—Energy consumption as an indicator of the wealth of nations[2,3]

To be sure, the situation has created nasty problems for large petroleum producing countries such as Indonesia, Venezuela and Nigeria, in addition to Russia, whose economies depend almost entirely upon oil (Figure 2). Their overextended credit in the period of the oil boom never allowed them to recover.

One of the most significant effects of low oil prices is that the energy industry has become uncharacteristically transparent, subtle, even taken for granted. This effect has contributed to a certain level of national bliss—one that subconsciously belies the international importance of the industry.

An interesting comparison can explain both this subtlety and the hidden strength of the petroleum industry. Consider the latest revenue totals of some important industries in the United States:

- Aerospace manufacturing and airlines—$250 billion
- Computers, including software and hardware—$420 billion
- Automobiles—$435 billion
- Petroleum and chemicals—$478 billion
- Banking and finance—$616 billion[7]

Similar industry groups and relative sizes are found in the European Union: Boeing is balanced by Airbus, General Motors by DaimlerChrysler, and ExxonMobil by Royal Dutch/Shell.

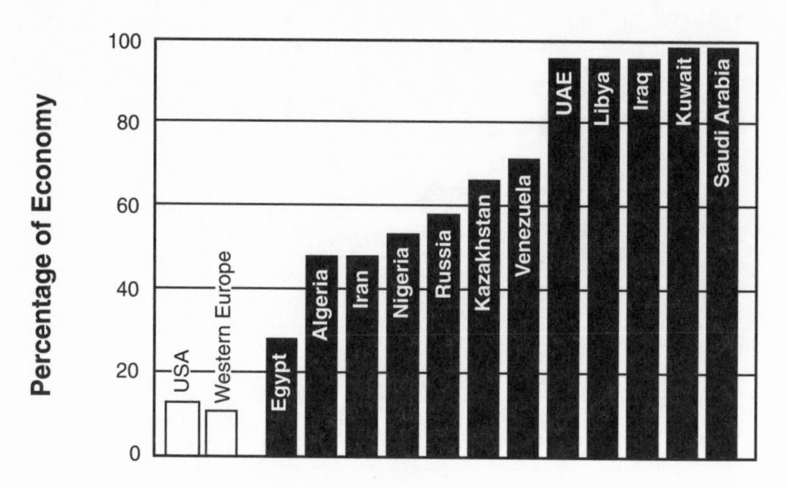

Fig. 2—Petroleum related fraction of selected economies[3-6]

Among Americans and western Europeans, the petroleum industry is viewed as one of many industries of similar magnitude, but this is a matter of limited local perspective. Although the rest of the world adds very little to banking, airplanes or computers, it adds enormously and almost exclusively to the petroleum industry, so much, in fact, that petroleum becomes larger than all of the other industries, combined. This is the relative and hidden power of the petroleum industry, and the reason for its solid future.

This industry—having shed 50,000 workers in 1998 and more than 1 million workers since the early 1980s, with the loss absorbed almost painlessly by the United States and the traditional oil producing states—is today as lean and mean as any industry can be.

A lean petroleum industry, armed with modern technology, may be the developed world's most potent asset today. The ExxonMobil and BP Amoco mega-mergers poise the industry for the next boom, which will surely include Saudi Arabia, Iran and the already reopened Venezuela.

The Wealth of the People in the Industry

Although not obvious to everyone, there is no doubt that the unbelievable riches of the richest corporations and sheikdoms spill down to ordinary workers and servants alike.

The current structural changes in the industry have been excruciating, with a lot of people falling by the wayside. However, for petroleum executives, independent producers, engineers, technicians and tobacco-chewing rig hands alike, it has been a "heckuva ride and it ain't done yet."

An entire generation of workers has retired from major oil companies during the past decade, many with "packages" that quietly made them multimillionaires. This often happened as early as the retiree's 50th birthday. In addition to a full retirement, they left with company stock, pay deferral and incentive investment funds that may have added 15 percent or more to their base salaries for decades. And this does not include the after-tax investments afforded by large salaries.

Salaries in the oil business have always been, and remain today, the highest in the world. In no other industry can a 22-year-old petroleum engineer, fresh out of school, take a first job at more than $50,000 per year—three times the salary of a schoolteacher, twice that of a starting civil engineer, and on par with or higher than that of most recent MBAs. Interestingly, the only youngsters making more money today are computer geeks like Bill Gates and Michael Dell, who dropped out of school before graduation—and there aren't very many of them (Figure 3 and Table 1).

How can one benefit from the biggest, most resilient, most successful industry in the world?

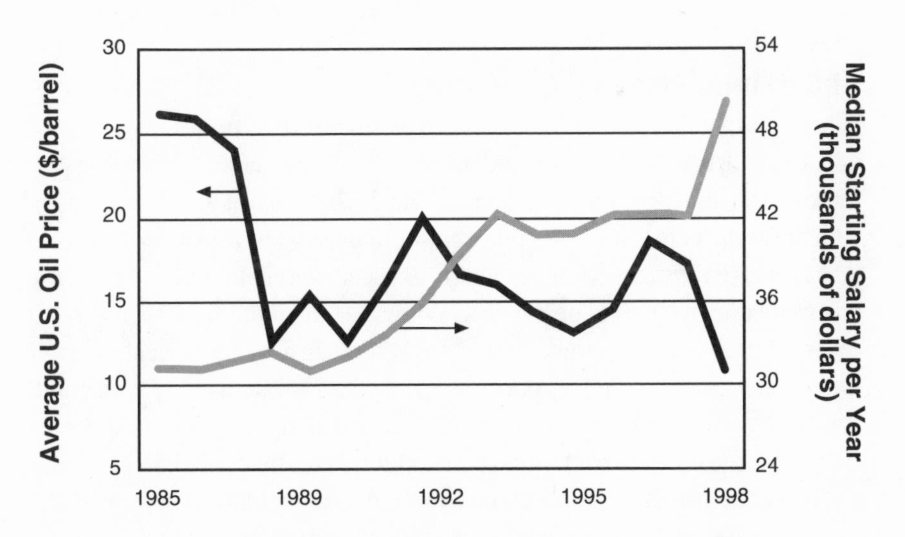

Fig. 3—Starting salaries with B.S. degree in petroleum engineering[8,9]

Table 1—Hot-Track Salaries[10]

Position	Annual Salary
Nanny	$24,000
Catering director	$30,000
Environmental accountant	$42,750
Speech pathologist	$44,000
Relationship manager	$48,300
Web site developer	$49,600
Molecular biologist	$50,000
Physical therapist	$57,200
Human resources director	$65,050
Construction manager	$65,064
Electrical engineer	$71,379
Network architect	$77,600
Gospel music artist	$85,000
Real-estate attorney	$95,924
Petroleum engineer	**$100,230**
Primary care physician	$128,000

Understanding the Big Picture

Since World War II, oil industry leaders have emphasized training over education, to the detriment of a true understanding of the issues involved. The ability to perform a 10-minute calculation on the back of an envelope, coupled with the capacity to think and understand, often has a literal worth many times that associated with the ability to run the most advanced model or other black-box simulator. In terms of personal economic benefit, it may be better to take a course in philosophy or symbolic algebra rather than a "results-oriented" training seminar, because it forces one to think. And it's always good to read and to read voraciously.

As Steve Forbes likes to quote, "With all thy getting, get understanding." Certainly the personal ability to view, anthologize and predict events in the world's biggest business represents tremendous personal value.

A big-picture view of the oil industry suggests an equilibrium oil price at the turn of the century of just under $20 per barrel, based on the traditionally inelastic demand and extreme elastic supply. Consideration of the financial, political and physical forces suggested that the 1998 sub-equilibrium oil prices needed to be balanced in 2000 by prices above the equilibrium; 1999 was a year of transition. Insiders of the world's biggest business understood this better than most.

At the time of the writing of this book, the world consumes 200 million barrels of oil per day equivalent. Of this, 40 percent is oil, 22 percent is gas, 24 percent is coal, 6 percent is nuclear, and 8 percent comprises all other energy forms, mostly hydroelectric.[11,12] (The renewables, wind, solar and the rest combined, comprise less than 0.5 percent.)

Contrary to popular notions, there was never really an oil glut at the end of the millennium. For a decade, world energy demand increased at 1.5 to 2 percent per year. In 1998, with many economies in recession, energy demand still increased, albeit only by 0.5 percent. Demand started to increase again in 1999, back in the range of 1 to 2 percent. In fact, oil consumption increased almost monotonically ever since the original Col. Drake well in 1859, with exceptions in 1974-75 and 1980-81.

The supply side of the equation is widely, and inappropriately, characterized by the Saudi cheap-oil-forever myth. This myth was firmly ensconced in the public discourse by a March 6, 1999 cover story in The Economist titled, "Drowning In Oil."[13] The article was laced with warnings that the Saudis may "throw open the taps," and "herald an era of $5 oil." The specter of oil selling for $2 or $3 per barrel was even raised.

Misplaced emphasis is constantly given to the low lifting costs of major national companies in Saudi Arabia, Venezuela and similar countries. The reality is that the up-front costs associated with activating these fields are some of the highest in the world. Development costs range from $3,500 per barrel per day of new production in Saudi Arabia and Venezuela, almost 3.5 times the costs in the U.S. Gulf of Mexico, to $7,500 per barrel per day in Iraq, to as high as $15,000 per barrel per day in Kuwait. (See *Part V—Primary Colors.*)

An abundance of prolific fields notwithstanding, it would take $100 billion and four to five years for Saudi Arabia to open its proverbial floodgates, something it can ill afford, either financially or politically.

An insider understands that it is constant and real intervention—people, drill bits and rigs—not market or political movements, that sustains production; without this intervention, production in every field in the world will decline. This is the simple physical law at the core of the petroleum business, a law that is poorly understood by most analysts. Without new capacity, global oil production would decline at 6 to 10 percent per year.

There was another tenet to support oil prices above the equilibrium in 2000: much of the international oil exploration and production machine was shut down from the fall of 1998 through most of 1999, with companies seeking relief through mergers and acquisitions, and restructuring for survival at $12 per barrel oil.

In many cases, CEOs were coerced by Wall Street to react to low oil prices; had they not, their stock price would have suffered. Royal Dutch, the only company among the Big Three that did not announce a mega-merger during 1998, incurred a precipitous drop in its stock price from more than $60 per share to $39.75 per share late in the year, while share prices at the other two big multinationals remained flat or increased slightly. Royal Dutch/Shell Chairman Mark Moody-Stuart responded by saying, "I am absolutely clear that our group's reputation with investors is on the line"[14] and that the company's upstream operations would "radically reorientate."[15] Six months later, Royal Dutch stock traded for more than $67 per share.

Was the company terribly mismanaged, bloated, and in need of dramatic cutbacks and reorganization? Probably not.

Sometimes, even experts in this industry utter complete nonsense. Right after an OPEC agreement to curtail oil production in March 1999, Daniel Yergin, Pulitzer Prize-winning author of "The Prize," said on CNN that he believed that OPEC was serious, and that the oil price would probably reach "$14 to $15 per barrel by the end of the year... if OPEC is disciplined." It reached that level in three days. In November 1999, with oil selling at more than $27, Yergin made a new prediction of "$30 or even higher by the end of the year."

The international petroleum industry has failed miserably to extricate a very weak link in its otherwise seamless integration: the speculators in commodity pricing and trading. On the positive side, this presents a clear opportunity for those who understand the fundamentals of the industry.

Investing in the Petroleum Industry

A repeatable 25-percent annual return on investment, built on an understanding of the physical, financial and political aspects of the petroleum business, is clearly a get-rich-slowly recipe.

For example, for those that recognize it, a tremendous investment opportunity underlies the dramatic, ongoing shift from oil to natural gas as the basic fuel of the U.S. economy. Already, the use of oil for space heating and electric power generation in the United States is well below 5 percent. Many pundits correctly point to the use of hydrogen and fuel cells for powering the next-generation automobile, but very few recognize or admit that natural gas will be the dominant source of hydrogen. So, maybe it is time to learn to trade natural gas futures, or even gas itself. Someone who understands the industry is far more capable of capitalizing on the value of natural gas than any MBA-trained "analyst."

In North America, a lot of incremental production can be brought on very quickly and with low capital requirements. The major consolidation and divestments by large oil producers in 1998 will, in turn, open a new wave of opportunities for small oil producers. While many are abandoning the oil industry as a dour investment, those who really understand the business are quietly reinvesting.

Those who talk endlessly about the price of gasoline at the pump show enormous ignorance of the petroleum industry. Pundits and journalists who agonized over gasoline at $1.50 per gallon in the United States in mid-1999 are the same people who, only a few months earlier at the time of the "oil glut," did not realize that $0.79 gasoline was a more striking aberration, especially when $4 gasoline is an accepted part of life elsewhere around the world. The pump price is now almost inconsequential compared to the cost of car lease payments, insurance, and maintenance, which combined, may be 10 times the cost of the fuel.

Now is the time to buy energy stocks. They will escalate in value substantially in the early 2000s. The wise investor buys for the long-term because energy is the world's biggest business, and it continues to move unstoppably forward.

Part II
Black

The substance of oil and the physics of finding and producing it

The first hydraulic fracture treatment, 1947.
(Courtesy: Halliburton Company)

In early 1983, I got a telephone call from a frantic public relations officer for the University of Alaska where I was teaching. NBC was doing a story on a scam in California, already in both criminal and legal litigations, and wanted to interview a neutral expert (one who did not work for the oil industry) on the subject.

A couple of con men had started a company in California by selling Alaskan oil lease rights to retirees and assorted small investors. In those younger days, I was immediately sure that the deal was a rat, and at the same time, I was filled with righteous indignation: How could they prey on unsuspecting little old ladies?

I knew Alaska's oil activity quite well: massive oil production at Prudhoe Bay and lesser, but quite involved, production offshore in the Kenai Peninsula. The very large multinational companies who ran these operations would have neither a need nor the inclination to find investors in California's little towns.

I returned the NBC producer's telephone call, and he explained the situation to me. It was exactly as I had suspected. The story involved drilled-and-abandoned wells that had reverted to the State. Some of these wells were in the most unlikely places. But, of course, this was Alaska, and if Prudhoe Bay could produce 1.5 million barrels per day, as the sales pitch implied, why wouldn't these wells make everybody else rich? The fact that the North Slope was 700 miles away didn't really figure into the argument.

The network flew me down to San Francisco. I was picked up at the airport and driven to the St. Francis Hotel, where the television crew had set up in one of the conference rooms. There, I met the newsman and his producer, and one of the victims of the scam, a slight 70-year-old woman. I have never been sure if the whole scene was natural or a prop, but I would not put it past the news folks: The old lady was in tennis shoes.

I was interviewed for a couple of hours. After we dispensed with the details of what all of us agreed was an obvious scam, the questions, driven by the newsman's notion of fairness and balance, turned to oil drilling and production. Maybe the con men were not as bad as we were implying. What if they really thought there was something there? After all, many wells were drilled.

Why do oil companies drill dry holes? Are they incompetent, or do they do it for tax breaks or some other secretive and conspiratorial motive? Why do they not put wells on line that the companies themselves declare have "shows of hydrocarbons"? Why do some wells produce so much more than others?

When it aired a few days later on NBC national news, my two-hour interview, much of it on the physics of oil exploration and production and the economics of exploitation, was reduced to four 15-second sound bites. – M.E.

✧

Oil is black because it is a mixture of many compounds with so much variation in physical properties that all light is absorbed. This blackness contrasts the crystalline transparency of water, the other main fluid that man mines from the earth, and contributes to the great mystery often associated with oil.

Everything that surrounds us—gases, liquids and solids—are compounds consisting of smaller parts, or elements. These building blocks are universal, not just on Earth.

Almost everything, from water to common salt to sand, consists of one element connected to one or two others. Oil, it turns out, is a bit more complicated.

In principle, *crude* oil, as it is often called, consists primarily of two elements, carbon and hydrogen. Yet, because carbon is one of few elements that can combine with itself, it forms long chains, branches and rings, each with its own properties. Thus, oil can be a mixture of literally hundreds of compounds. Each crude is a unique mixture; collectively, these mixtures are termed hydrocarbons.

Oil, a word that is a direct translation from the Greek *petroleum* (*rock oil*), got its name from the ancient observation of black liquid naturally coming from the ground to the surface. To this day, by far the greatest portion of oil found on the surface of the earth—including on otherwise pristine beaches—seeps naturally from the ground.

Yet, the notion that oil is only a liquid is misleading, considering the makeup of petroleum underground. The sizes of the compounds, in conjunction with pressure and temperature, are the factors that divide petroleum into oil and gas. Small molecules, the smallest of which is methane, are likely to be in the gaseous phase within a great range of pressures and temperatures; longer molecules remain in liquid form.

The depth at which petroleum is found correlates well to pressure and temperature, in spite of occasional anomalies such as high temperature or large pockets of pressure near the ground's surface. The reservoir pressure and temperature dictate both the composition and the in-situ phase separation of petroleum into oil and gas. At depths of 3.5 miles (6 kilometers) or more, generally only gas is found. At shallow depths of 0.5 miles (1 kilometer) or less, petroleum is likely to be only liquid. Between these two extremes, where the great majority of petroleum reservoirs are found, oil and gas coexist, and their relative volumes (for a given earth temperature) depend upon the pressure.

The sizes of the molecules in a particular petroleum mixture not only define the makeup of the gas and liquid phases, but also give oil its flowing character. Smaller molecules result in a very low-viscosity fluid; large molecules increase the viscosity. In some cases, viscosity is increased to the point that a crude, placed in a beaker at room temperature and turned upside down, does not pour out or even deform.

Refineries separate the crude petroleum into individual components or groups of components of similar properties and potential uses. For example, butane is a pure component that is extracted in a refinery. Other refinery cuts are mixtures of compounds that form jet fuel, automobile gasoline and heating fuel.

Dissatisfied with the naturally occurring crude compositions, refiners "reform" the smaller molecules or "crack" the larger molecules to increase the yield of more valuable and useful compounds.

Refinery outputs are also supplied as feedstock to the petrochemical industry, which depends almost entirely upon petroleum. Petroleum-based synthetic materials such as polymers, plastics and fibers exist in virtually every modern industry, from food packaging to thread for clothing to computer casings and parts and everything in-between.

Of course, the most discernible and commonly understood everyday use for petroleum is energy conversion—powering transportation, generating electricity, and providing direct heating of spaces.

The conversion of petroleum (oil or gas) to energy is a result of combustion, the reaction of hydrocarbons with oxygen. The release of energy is the most important characteristic of this reaction. There are also chemical products of the reaction, the most common being water and carbon dioxide. The increase of the latter in the atmosphere has been associated with one of the scariest and most widespread topics of both political sloganeering and scientific research: the greenhouse effect.

The Origin and Migration of Petroleum

How was petroleum created and why is it found in much greater abundance in certain parts of the world? More to the point, why are some areas that are close to prolific petroleum producing areas largely devoid of it?

Fanciful and even extraordinarily incredible theories have surfaced over the years about the origin of oil-field hydrocarbons. The one most accepted, although not quite reproducible in all its details either in the laboratory or in modern computer simulations, suggests that petroleum derives from organic (originally living) matter that has been deposited and buried for tens- to hundreds-of-millions of years. One of these geologic eras, widely known because of recent popular films, is the Jurassic.

Envision ancient river deltas and ancient coasts with large accumulations of sediments. Envision these features, in turn, being covered with many more layers of sediments, not necessarily bearing the same amounts of organic matter, but together adding the necessary pressures and temperatures by virtue of their overburden weight and associated depth. These conditions have generally been postulated as requirements for transformation of the buried organic matter into petroleum hydrocarbons.

More recent geologic events, including earth plate tectonic movements and the widely accepted migration of the continents may have changed the position of these ancient coastlines. Some of them have been uplifted and are now thousands of kilometers inland. Others have sunk into much deeper waters; in some cases, 3 kilometers or deeper. Even more interesting phenomena may have occurred, however; ancient petroleum provinces may have been split in conjunction with the drifting of continents, and may have drifted away. Geologic formations found today in West Africa have sister formations on the eastern coasts of the United States and South America.

Such is a first pass in understanding petroleum geology.

Taking a clue from modern coastlines where river deltas and their sizes and distributions vary considerably, it is not a stretch to postulate that petroleum formations must have some unique features.

First, although the organic origin of petroleum can be accepted readily, the formations that bear it are not evenly spread. Furthermore, petroleum is not necessarily found where it was created, but instead, may have migrated from its source rock and accumulated in a reservoir rock. The distance between the two could be tens of kilometers.

Because petroleum generally has a much lower density than water, which is always present (evidence of the marine origin of petroleum), the migration is always upward, with oil sitting on top of water. If gas is present, it will be at the very top, above oil.

The rocks that hold petroleum are also special.

There are no pools, caverns or underground rivers of oil, although the popular press often alludes to them. When examined under a microscope, the rock appears to consist of grains and interconnected pores. Petroleum inhabits these pore spaces of the rock. If the rock's porosity (a measurable quantity denoting the portion of the rock that consists of pores) is not large enough, the volume of petroleum present is not commercially interesting.

Sands are some of the most common reservoir rocks, accumulated from mountains in alluvial fans and rivers; in dunes on lakes and seashores; in tidal flats and bars in shallow marine environments; and in turbidite fans near the continental shelf.

Carbonate rocks, such as limestone and dolomite, accumulate in reefs and other marine settings and constitute the other main family of petroleum-bearing geologic formations.

The chronology of geological deposition is not only interesting in identifying layers and their depths; it also suggests important characteristics of the rock. For example, in a given location, a Devonian deposit will be deeper than a Jurassic, and it will probably be more compacted, tighter and less porous. In another location, these old rocks may be found on the earth's surface in a road cut or a cliff, along a beach or in a mountainous region. Even older rocks that are uplifted and outcropping on the surface have these characteristics, and the outcrops provide a tangible but indirect means to study the buried rocks.

Although petroleum tends to move upward, floating on water and finding its way through layers of porous rock, it normally gets trapped en route by a layer of rock called a cap rock. The cap rock has essentially no porosity, so it effectively blocks or "caps" the upward migration of petroleum and creates a reservoir. (Migrating petroleum that is not trapped underground manifests itself as a surface oil seep.)

The weight of the overlying or "overburden" rock causes the buried fluids to be under pressure, at times considerably higher pressure than the hydrostatic pressure (i.e., the pressure that a column of water of an equivalent height would exert at its base).

This is the nature of a petroleum reservoir, a porous rock containing fluid under pressure, awaiting discovery.

Exploration

Looking for and finding petroleum (the art of exploration) has always been shrouded with a healthy dose of awe, not just from the general public, but also from inside the industry itself.

It is perhaps because of the riches associated with a good hit and "striking it rich." It is because of the glamour and romance of remote and exotic places spanning the globe. It is because, to this day, in spite of enormous advances, the technology is still inexact and its results uncertain. It is because nature often plays games, taking pleasure in disappointments. It is because a good modern geoscientist is a rare individual, an almost mystical combination of intuition, education and good fortune.

First, the general geology must be right. In the early times of petroleum activity, some of the most obvious places to look for oil were spots with natural seeps. As theories of the origin of petroleum emerged, sedimentary basins became a prime target for the hunters. These basins are not so difficult to spot from geological outcrops and other features. However, not all petroleum reservoirs have obvious surface manifestations, so geologists use a lot of inferences and conjectures, often quite intelligently and sometimes not.

Until the 1980s, drilling was a hit-or-miss operation. New wells, even in presumably prolific areas, were termed "wildcat," and people and companies would brag or lament of their success or failure, depending on how they bobbed above or below 10 percent (i.e., one good well and nine dry holes for every 10 wells drilled).

Few technologies in the history of the petroleum industry can match the importance of seismic measurements and their impact on exploration and, eventually, production.

Seismic measurements begin with the generation of a seismic event, a mini-earthquake that is transmitted downward from the surface. In the early days of the technology, explosions were used. Today, in offshore locations, a specially designed vessel with side-guns shoots large air bubbles into the water, creating a concussion that hits the sea floor and produces a local vibration. On land, heavy-duty thumper trucks create vibrations by hammering the ground.

In all cases, a seismic wave migrates into the ground, traversing layers (strata) to depths of 3 kilometers or more. The signal en route reflects from certain layers and bounces back. As it crosses strata, the wave undergoes refraction, the same phenomenon that causes a pencil to appear broken when it is put into a glass of liquid. Just as different liquids give different broken-pencil angles, different geologic strata provide different refraction effects as the seismic wave traverses them. On the surface, these wave responses are detected by a line of receivers.

The massive data points that are collected are then visualized and analyzed. This exercise is so computer-intensive that, for years, the petroleum industry was by far the biggest user of super-computers. During the recession of the late 1970s, the oil and gas industry was key in the survival of the makers of these machines.

As the seismic vessel or thumper trucks move with their detector arrays—easily accomplished at sea, but more difficult on land—successive seismic data planes are overlain to produce a 3-D image, or cube (Figure 1).

Seismic measurements have evolved well beyond their traditional and most obvious role as an exploration tool. Today, seismic attributes are integrated with many other measurements to provide new insights, such as the detection of gas and the movement of fluids within the reservoir. Performed in time-lapse sequences with production—a process called 4-D, with time as the fourth dimension—seismic measurements have become an equally important production engineering tool.

This technology has been the mainstay of successful re-exploration activities around the world.

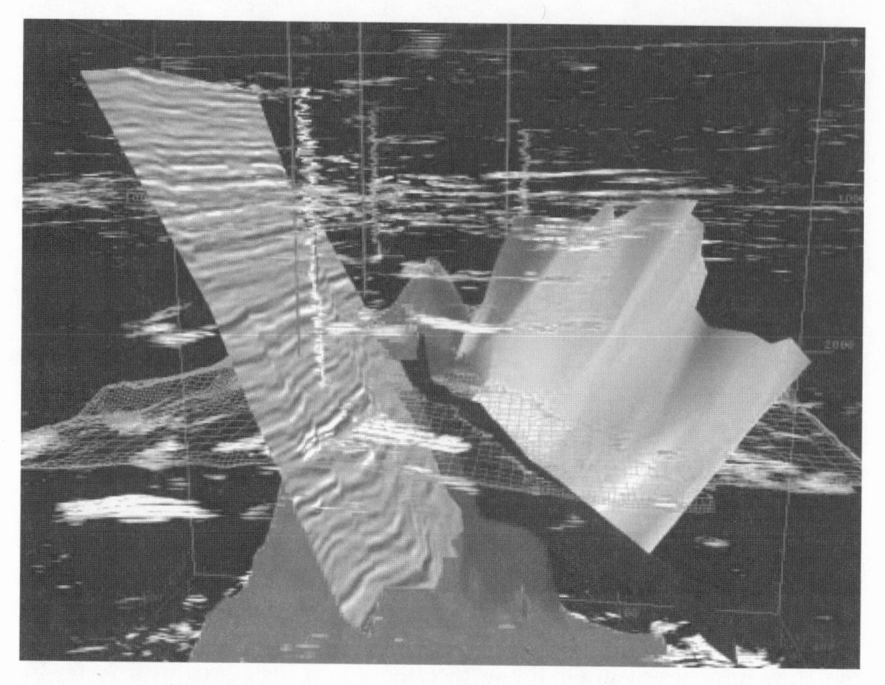

Fig. 1—Seismic visualization shows characteristics of geologic formations. (Courtesy: Landmark Graphics Corporation)

The future of the petroleum industry lies not only in the discovery of massive new reservoirs. With the exception of the deep offshore, few reservoirs like Alaska's Prudhoe Bay are left to be found, and for certain, not another Middle East. Future production is more likely to come through the use of new technology and re-exploration techniques in exploiting known petroleum provinces to locate bypassed oil or oil that was previously considered marginally economic.

Yet, no formation evaluation technology can substitute for the intense business of drilling wells to tap the petroleum resource.

Drilling and Well Construction

The idea of drilling a hole in the ground to find water has been around for hundreds of years. The Chinese claim to have done it for thousands of years.

In 1859, "Col." Edwin L. Drake drilled the first purposeful oil well, in Titusville, Pennsylvania, all of 69 feet deep. In one of the first examples of the "can do" attitude that has stayed with the industry ever since, Drake and a blacksmith cobbled together existing salt-mining tools with a steam engine (to drive the drill bit to greater depths) on an intermittent $1,000s budget supplied by less-than-convinced investors—and struck oil![1]

As practiced today, drilling is an involved operation with heavy-duty equipment, a variety of fluids and sophisticated instrumentation.

Rotary drilling has been around for many decades, using one of the most identifiable symbols of the oil field: the drilling rig. At the end of the drillpipe, the drill bit grinds away at the rock while it is lubricated by the drilling fluid, which is known not-so-disparagingly as "mud." Mud is actually a rather complex fluid, with lots of additives for effecting desired functions.

Depths vary with the position of the targeted geological strata. Oil producing formations are typically found at depths of 0.5 to 2.5 miles (1 to 4 kilometers); gas reservoirs are often 3.5 miles (6 kilometers) deep or deeper. The world record was set when an experimental research well in Germany was drilled more than 9 miles (15 kilometers) deep.

When a well is drilled from an offshore platform, the drill bit may traverse hundreds of meters of water before hitting solid ground. It is very likely that wells will be drilled in water depths of 2 miles (3 kilometers) or more in the near future.

The size and technology of modern drilling operations are greatly advanced compared to the original Drake well in 1859, yet basic similarities remain between the derrick and associated equipment (Figure 2).

Because of repeated oil boom-and-bust cycles, the number of wells drilled annually fluctuates wildly—between 40,000 and 80,000 per year since 1980. About half of these wells are drilled in North America, and the other half, in the rest of the world.

The cost of drilling always represents a respectable chunk of petroleum operations (50 to 60 percent of a company's development budget), but the numbers range widely. Based on the economy of scale that comes from drilling many wells, a 2-mile (3-kilometer) deep well in the United States can be drilled for 10 percent or less of the cost of drilling the same well in many parts of South America, the Middle East and Africa. When all ancillary items are considered, offshore exploration wells may cost four times more than the most expensive land wells.

Fig. 2—Similarities between the Drake well (1859) and a modern offshore drilling operation in the North Sea (1997) are striking.

Advances in hardware and measurement techniques have made it possible to drill highly-deviated and horizontal wells for maintaining the well trajectory inside an oil-bearing formation. Recently, even more complex well architectures have been developed for drilling multilateral, multilevel and multibranched wells. Both well deviation and the creation of more branches serve to increase contact between the well and the formation and ordinarily result in a much higher production rate. An array of high-tech tools and instrumentation now makes it possible to take measurements-while-drilling and to track the advance of the drill bit in real time. This gives rise to what has been termed "geosteering," or the guiding of the drilling process inside a target formation.

Once a target formation has been identified and a well drilled, the well's petroleum potential must be evaluated. If the well holds sufficient potential, it is prepared for production, or "completed."

In the early 1920s, two French brothers, Conrad and Marcel Schlumberger, started with a simple theory for evaluating a well's potential. They experimented first in a laboratory at the Ecole des Mines de Paris and then performed several field trials in France and the United States.[2]

The principle was simple, yet ingenious. If one were to accept the notion of fluids in porous media, and if two kinds of fluids could be envisioned—useless water and useful petroleum—the electrical properties of the fluid-bearing formations would be drastically different. Oil, being a poor conductor of electricity, would cause a high degree of resistance or resistivity (resistance per unit area). Conversely, water in the form of brines would create very little resistivity. The Schlumberger brothers showed that electrical measurements taken in a well could be used to precisely identify rocks with good petroleum potential while ferreting out less attractive rock layers.

The brothers' big break came in 1931, following a test well in Grozny at the foot of the Caucasus. Data produced with their invention correlated with definitive proof of the presence of oil.[2]

Thus, well logging was born. Starting with a contract in the Soviet Union and eventually spawning an entire area of petroleum science and proficiency, logging involved not just electrical resistivity, but a whole suite of other measurements and instruments with which to collect them.

Logging also made the Schlumberger name one of the most prominent in the business, and one that is still carried by one of the largest and most technologically advanced companies in the field.

A well, if left as an open hole, quickly develops problems. First, in unconsolidated or weak rocks, the well may fill up with debris from the reservoir or collapse outright. Even more worrisome, the well could become a conduit for communication between different petroleum reservoirs whose commingling is not desirable. In the worst case scenario, petroleum fluids and drinking water aquifers could commingle. Commingling not only refers to the mixing of physical properties of different fluids, a definite concern, but also to the equalizing of fluid pressures between different formations. Fluids produced from higher-pressure reservoirs that flow through the well into lower-pressure formations will not reach the surface.

The solution is to line, or case, the well with a metal pipe. The annulus between the formation and the pipe is filled with a cement slurry that, when set, provides a seal that isolates disparate formations. This technique—pioneered in 1920 by an 18-year-old entrepreneur named Erle P. Halliburton, in a well owned by William G. Skelly near Wilson, Oklahoma—not only caught on, but also led to the creation of another large service company that still bears the inventor's name.[3]

Of course, a cased and cemented well still needs a path through which petroleum can flow from the identified desirable formations. Therefore, explosive charges are used to create holes, or perforations, in the pipe. If the cementing provides the seal and isolation, and if perforating is successful and on target, the well is appropriately constructed.

Production and Decline

Reservoir pressure drives fluids into the well. Although porosity, along with the reservoir thickness and areal extent, indicates how much petroleum is in place, another property—permeability—dictates how fast the well can produce the fluid. This is the well production rate. Given the same initial pressures in two reservoirs of the same thickness, a reservoir whose permeability is 100 times the permeability of the second reservoir will produce at a rate that is 100 times higher.

Because fluid converges from a large drainage area into a well, the condition of the near-well zone becomes crucial. The permeability of this zone is frequently damaged during drilling and well construction. In turn, this damage can severely limit production from the well.

Both man-made damage and low initial permeability can be rectified with production enhancement or stimulation. Damaged wells can be restored to their full, undamaged producing rates. Most wells, low-permeability wells in particular, can be stimulated to produce at rates that are even higher than their undamaged producing rates, though the costs and associated benefits vary greatly. One option for stimulating production is to remove the near-well problems with chemicals that dissolve or bypass the damage. Although such well treatments are not always successful, they have been widely practiced, with enormous benefits.

Another frequently used technique is hydraulic fracturing—injecting highly pressurized fluids at a very high rate to create a crack in the reservoir. The fracture may be propped open with millions of pounds of clean, uniform sand, resulting in a permeability that is orders-of-magnitude larger than the surrounding reservoir. The injected sand creates something equivalent to a superhighway, and fluids moving along this highway can be produced at much higher rates.

Deviated wells, horizontal wells and the recently introduced multibranched wells are also examples of production-enhancement tools that have the potential to produce at much higher rates than single vertical wells.

Production, either slow or accelerated, always becomes self-defeating as time advances, however. Production rate decline affects all petroleum wells.

Underground fluid withdrawal brings a decline in the reservoir pressure, analogous to the emptying of a balloon full of air. Except, unlike a balloon, the total porous rock volume does not shrink in size; the remaining fluid expands to occupy all available space.

Expansion of reservoir fluids is perhaps the most easily understood mechanism that drives petroleum production. The volume of a liquid does not change much when it is pressurized or depressurized; gas is far more compressible. Therefore, if the reservoir fluid is only liquid, relatively small fluid withdrawals can deplete the pressure rapidly. Recovery of 3 percent or less of the initial oil in place can make the reservoir pressure equal to the pressure at the bottom of the hole. When this occurs, fluids are no longer driven into the well and 97 percent of the original oil is left "in place" in the reservoir. This defines *primary recovery*, the most elemental but generally unacceptable ending point in petroleum exploitation.

It is important to remember that the reservoir pressure is not only the driving force behind fluid movement, but also the main influence on fluid phase behavior.

A specific (lower) pressure level called the "bubble-point" pressure marks the onset of natural gas evolution, known as solution gas. When this level is reached (the point at which this occurs depends on the specific crude), recovery can increase substantially to 15 percent or more.

If a large water aquifer is in contact with the petroleum reservoir, a natural drive mechanism can be provided by natural water influx. The larger the aquifer, the more effective and the more long-lived this drive mechanism tends to be. If a strong water drive is in effect, 10 to 25 percent of the oil in place can be recovered.

In theory, the total production rate in a petroleum reservoir follows a bell-shaped curve, as suggested by M. King Hubbert, one of the great names in petroleum engineering.[4] The idea is that a fully developed field will reach a maximum output level, which marks the beginning of reservoir maturity, and then begin its decline. This character is suggested by the composite of production from all Omani fields, which peaked in 1997 (Figure 3).

As reservoir pressure declines—which will always happen, though the rate of decline may vary—the fluid must be brought to the top of the well with some imposed lift mechanism. Gas lift reduces the density of the fluid in the well by injecting gas from the top and mixing it with the produced fluid at the bottom. This reduces the hydrostatic head pressure that must be overcome for the well to flow.

Another common technique is to pump out the reservoir fluid physically. By far the most common sight in oil fields around the world is the "beam pump," so called because the downhole pump and connecting rods are actuated by a reciprocating steel beam at the surface—the walking beam.

Reservoir water eventually and commonly encroaches into wells from an underlying aquifer through a phenomenon called coning. In this situation, reservoir water becomes an overwhelmingly parasitic fluid. The most expensive activity in petroleum production is the separation and disposal of produced water. Disposal is sometimes accomplished by evaporation, but more commonly, by reinjection into some other underground reservoir.

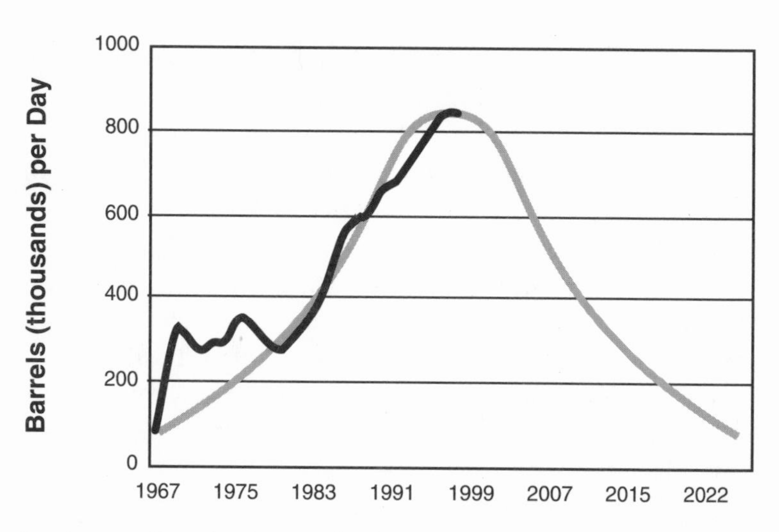

Fig. 3—Oman oil production overlaid on classic King Hubbert bell-shaped curve[4,5]

This is not the only problem caused by water. The most insidious effect is that reservoir water competes with petroleum for the flow paths near the well, dramatically reducing the amount of oil produced.

Maturity or "old age" in the oil field means any or all of three things: reservoir pressure depletion, the existence of a much larger water fraction in the production, and production from lower-permeability reservoirs. The latter is a result of drilling into progressively lower-quality geological structures after the choice targets in an area have been exploited. In all cases, maturity is marked by far lower petroleum production rates and substantially higher lifting costs per barrel of oil produced.

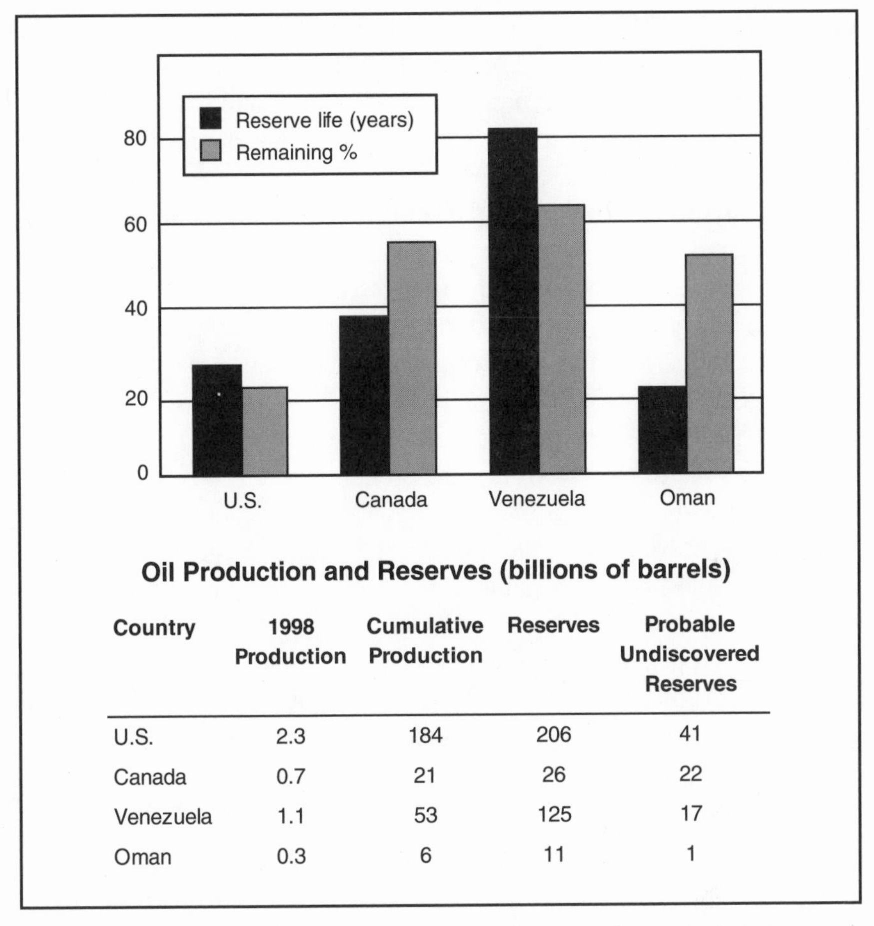

Oil Production and Reserves (billions of barrels)

Country	1998 Production	Cumulative Production	Reserves	Probable Undiscovered Reserves
U.S.	2.3	184	206	41
Canada	0.7	21	26	22
Venezuela	1.1	53	125	17
Oman	0.3	6	11	1

Fig. 4—Significant amounts of oil remain in the ground even in the most mature and intensely produced theatres of operation.[6]

It is not that divine providence has bestowed more prolific reservoirs upon the Middle East. Wells in Oklahoma and Texas produced at similar rates, except for one thing: They did it 60 or 70 years ago.

Many methods for sweeping petroleum through the reservoir have been tried. Each involves the injection of fluids in strategically placed wells to push petroleum into producer wells. The most basic idea is the injection of water—in many cases, produced water that has been recycled and treated. This process, known as waterflooding, or *secondary recovery*, has been applied throughout the world, and is still widely used today. Techniques for attacking high viscosity and other impediments to flow are collectively known as enhanced oil recovery (EOR) processes. Many of the latter can be expensive and uneconomical (i.e., their cost cannot be offset by incremental production at a given oil price).

No matter what, petroleum exploitation involving any natural or human-assisted recovery still leaves a large percentage of the original oil in place. Even in the most mature and intensely produced reservoirs, far more oil may remain than has been produced (Figure 4). Production of any incremental oil becomes an excruciating exercise that requires technology and sound management, and depends greatly on the economics of the day.

What is certain is that the life of each reservoir and even entire petroleum provinces is predictable, from young and prolific to old and difficult. Economic considerations always favor accelerated production. The concept of net present value makes this true in all industries—i.e., the sooner one gets the money, the better. It is made doubly true in the oil industry by a revenue stream (production rate) that naturally declines with time.

Prolonging petroleum production for a considerable time, once decline ensues, is both possible and common. Hubbert's theoretical distribution of production (referred to as "log-normal") has been replaced in practice by the

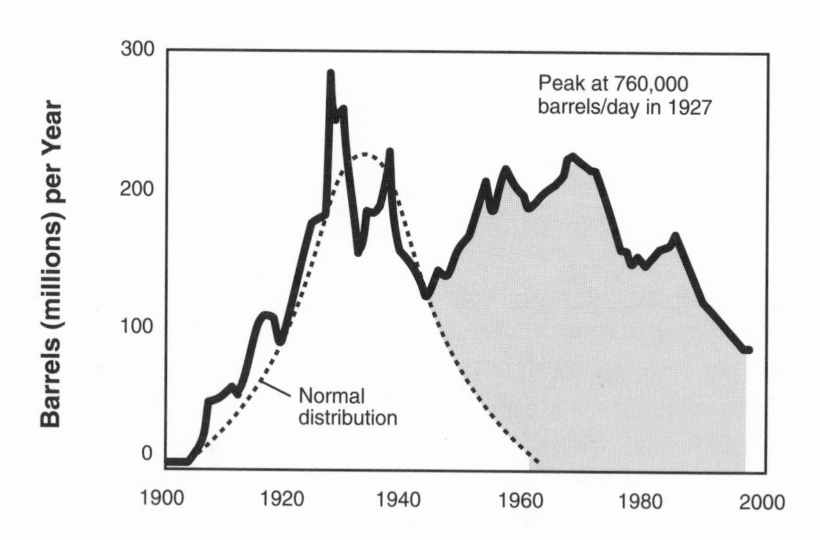

Fig. 5—"Fractal" distribution of production in Oklahoma[8]

"fractal" distribution.[7] Late-time production can be extended through the exploitation of a very large number of smaller reservoirs and geologic features that are often overlooked or discounted in the initial, large-scale survey of a petroleum producing province.

The history of production in Oklahoma, an obviously mature area, provides an example of this fractal distribution (Figure 5).

This "second wind" has added 7 billion extra barrels to Oklahoma's cumulative production since its peak in 1927. It came with increasing difficulty, innovation and enterprise, and under very tight economic constraints. A single good well in Saudi Arabia today—or a single good well in Oklahoma 70 years ago—can produce as much as 1,000 active wells in the state today.

Part III
Red, White and Blue

Origins of the most American industry and the ghost of Rockefeller

Rockefeller on the way to court for the landmark federal antitrust suit.
(Courtesy: Rockefeller Archive Center)

I rumble through town in a brand new Ford F350 crew cab diesel pickup, past countless gray crumbling buildings and unlit courtyards. Scattered kids, early in the street, stare and wave at the alien "masheé-na" (the only other F350 these kids have ever seen is driven by my driller, "Mo," from Midland, Texas).

This day in late 1993 finds me, a transplanted Alaskan, now living in Houston, halfway around the world in the former Soviet republic of Kazakhstan. I am representing the New York firm of Edward Carey, brother of former New York Governor Hugh Carey, who has negotiated one of the first petroleum joint ventures in the Caspian region. Millions of dollars changed hands, and a dedicated shipload of drilling equipment has already made it from Houston to the little desert town of Aktau.

People have called this place the Midland, Texas, of Kazakhstan, the next oil boomtown, but being here, it is difficult to believe that anything too significant is about to happen. The water in my flat runs hot-only on some days and cold-only on others. (Cold-only days are better, because you can at least take a shower; a shower with 150 F water is not even an option.) Without the infrastructure to take a shower, I think to myself, how are we going to develop an entire oil field in this place?

Oddly, the none-too-subtle KGB emissary who has been assigned to our joint venture, a stoic Russian petroleum engineer with no decipherable past, still describes this broken-down town as a great feat of the communist machine.

The specially designated joint venture building is a refurbished showpiece for the newly independent republic in its move toward capitalism. By the time I arrive, all 35-plus office employees who have been picked up by the JV bus (a 1930s-vintage converted schoolbus) will already be in their places, pretending to work. Pretending to work is a hallmark of the Russian presence here. So are the toilets that do not work. This part of the unspoken workday ritual calls for using the toilet before the bus picks you up in the morning, and not drinking coffee until almost noon … allowing you to do your necessary deeds over lunchtime.

I greet several people, "dohbro-ye ootra" (good morning), while searching for an interpreter, and my mind quickly focuses on the task at hand: I will spend most of the day reviewing our "joint" field development plan with Nurzan Djumagulov, the young and aspiring deputy director of the Kazakh-American Joint Venture, before heading on to the capital city of Almaty to seek final approval from the Ministry of Oil and Gas.

Even if we represent very different interests and cultures, Nurzan and I have developed a strong personal and professional rapport, so I expect a rather perfunctory review. Yet, by midday, I begin to sense a noticeable resistance. This strikes me as odd, given the day-to-day work we have done for months, explaining the plan to a seemingly endless array of local and regional authorities.

Eventually, I ask the interpreter to find out, off the record, what is bothering Nurzan. After a lively 10-minute exchange that I do not understand, the interpreter, a young Russian woman, turns to me and says, "He likes dealing with Americans, but you do things too fast." Nurzan was only 40 years old, and our 15-year development plan was too short, by about 10 years. In his words, "I am a young man…"

At once, it dawns on me that the time value of money has no meaning to a Kazakh. Once the joint venture development was complete, in Nurzan's mind, his job was over. The fact that our joint venture, if successful, would produce $1 billion in profit did not figure in.

So I spend the next 1½ hours trying, in many ways unsuccessfully, to explain that if we had the $1 billion today, we could develop many more oil fields, employing many more people. Or, if we developed all of the oil, many billions later, we could build a Caspian seafront resort or a theme park.

I look my young colleague in the eye and say, "How would you like to run Kazakhstan's Disneyland after we finish this project?"

Intellectually, he accepted that the Yankees were, in fact, the long-awaited saviors of his country; but this mixed uncomfortably with the sense that the locals were still the pawns in this scheme, somehow being taken advantage of and powerless to do anything about it. – R.O.

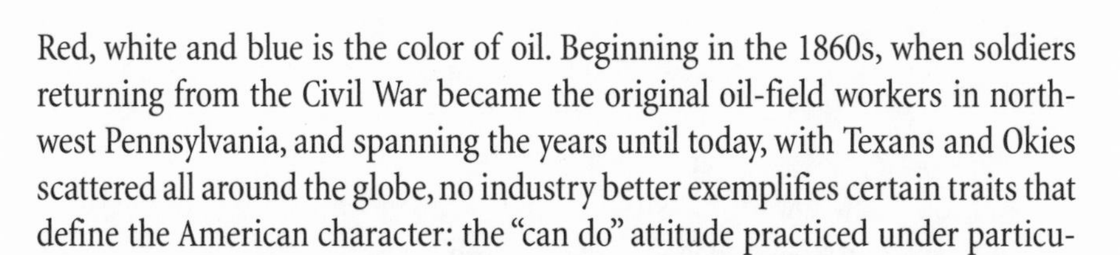

Red, white and blue is the color of oil. Beginning in the 1860s, when soldiers returning from the Civil War became the original oil-field workers in northwest Pennsylvania, and spanning the years until today, with Texans and Okies scattered all around the globe, no industry better exemplifies certain traits that define the American character: the "can do" attitude practiced under particularly hostile conditions; the piecing together of apparently irreconcilable geographic, political, financial and technical elements; and the frequent overexploitation of relationships and opportunities with abandon—a kind of *forced serendipity.*

The industry's toughness is often credited to the ruthless and enduring boom-and-bust cycles, the first of which was manifested within 24 months of the original Pennsylvania discovery by Col. Drake in 1859. Supplies of the turpentine-derivative illuminant camphene were interrupted by the Civil War. This gave the new industry a ready-made market for kerosene that was extracted from Pennsylvania crude oil. A swarm of oil refiners set up almost overnight around Cleveland and Pittsburgh. The kerosene cut alone provided the basis for an oil price of $15 per barrel almost immediately. (Gasoline was a waste product before Henry Ford.) Yet, by the fall of 1861, following a reckless race for production under the "rule of capture" and the 3,000 barrel per day Empire Well, the price plunged to 10 cents per barrel.

Riches implied by an oil "gusher" and the associated high-volume, low-cost production are predictably swallowed up by oversupply and depressed prices in an industry where "glut" is defined by as little as 2 or 3 percent excess capacity. Ready supply and low prices, in turn, promote increased use and demand. This sequence applied to kerosene in the late 1800s, and has applied to gasoline in successive waves in the early and mid-1900s. At the same time, low prices reduce the incentive to find new oil, and production from existing fields only declines, so every glut dries up and invariably the next oil shortage emerges. This puts upward pressure on prices, and the specter of riches re-emerges. So, the cycle repeats itself.

This boom-and-bust cycle has caused extreme hardship for oil producers throughout history, but it has also impeded the development of oil markets at various junctions as consumers, the public, industry and, most of all, government worried that oil might "run out."

In the 1860s, amid the maelstrom and unrestricted atmosphere of a virgin industry, a godfather emerged to embrace oil and exploit it to his advantage. Of course, that man was John D. Rockefeller. It is said that he never lost money in a single year, even through boom-and-bust cycles. Way beyond oil, Rockefeller created American-style capitalism.

Ron Chernow, author of "Titan," perhaps the most balanced and exhaustive biography on Rockefeller, says that his subject was "American to the marrow."[1]

America—Center of the World's Biggest Business

It is fair to wonder how an industry that is so eminently global—a quarter of U.S. kerosene production was shipped overseas almost from the very start—can be considered quintessentially American. What differentiates Shell and BP, products of European colonial pasts that are in frequent, indistinguishable symbiosis with their parent governments, from the U.S. flagship, Standard Oil, and its modern reincarnation, Exxon?

The advent of mass production in the U.S. in the 1860s and especially the imposing presence of a Rockefeller are important elements. However, a more compelling reason for America's dominance in oil is illustrated by an event that happened 80 years later.

Buried somewhere in the middle of Daniel Yergin's 1991 masterful narrative of the oil industry, "The Prize," a key event is chronicled under the heading "We're Running Out of Oil!"[2]—a proclamation made in 1943 by none other than then-U.S. Interior Secretary Harold Ickes. Concern mounted as oil was shaping up to be the key shortage for the Axis Powers, and at a time when the United States (the only nation with production in excess of its domestic needs) was almost single-handedly fueling the Allied war machine.

Ickes went on, "America's crown, symbolizing supremacy as the oil empire of the world, is sliding down over one eye." His implied conclusion was that the United States, and in particular, its government, should secure foreign oil reserves. The most obvious turn would be for the United States to look to the Arabian Peninsula, as Britain had since the end of World War I.

Herbert Feis, the State Department's economic adviser, said in 1943, "In all surveys of the situation, the pencil came to an awed pause at one point and place—the Middle East."[2]

Thinking of the British government's ownership in Anglo-Iranian Oil, Ickes proposed that the U.S. government take a 51 percent share in activities of the two private U.S. companies that were developing Arabian oil at the time, Socal (Standard Oil Co. of California) and Texaco. The enterprise would be named American-Arabian Oil Company.

Socal and Texaco were interested and willing, but the rest of the industry was fervently opposed to Ickes' proposal, fearing everything from curbed opportunities in the Middle East to subsidized government competition and federal control of the oil industry. The extreme would be nationalization.

Other majors, Standard Oil Co. of New Jersey (now Exxon) and Socony-Vacuum (Mobil), in particular, were also quite interested in gaining a foothold in Saudi oil.

Thus, in 1943, by all measures except for a fluke, the U.S. government could have owned the largest of all national oil companies. The implications could have been enormous:

- Had the U.S. government been the concessionaire—instead of Exxon, Mobil, Chevron and Texaco—Saudi Arabia might not have nationalized its oil industry in 1976.
- If Exxon's eventual role had been practiced by the U.S. government, the evolution of competing companies would have been dramatically different, both domestically and globally.

As it happened, the private U.S. companies were left to negotiate and self-finance the development of deserts in Saudi Arabia and other far reaches of the globe. This is what defined the American oil business.

Though the federal government did not own oil reserves directly, it still felt a need to be involved in the industry; regulations became the proxy.

While private industry had a close brush with major government involvement in 1943, controls by a much younger U.S. government were not even on the horizon at the start of the industry in 1860. Although the story has been told the other way around, it was the creative psychology and dynamics of a young country in the post-Civil War industrial boom that gave rise to a man named Rockefeller.

It was Rockefeller who was to become Exxon, with the blessing and often the active participation (witting or unwitting) of the U.S. government.

His huge imprint on Exxon and the modern U.S. corporation is manifest by contemporary business concepts such as standardization, distribution, vertical integration, research and technology, waste minimization and even corporate public relations.

The Rockefeller Story

Thousands of pages have been written by, on and about Rockefeller,[3-6] and it is still a formidable task to tell the story in a reasonable space without digressing to esoteric or metaphysical discussions of wealth and wealth-building that are so common in the popular press today. Another pitfall of recent Rockefeller biographers is their near-obsession with the post-modern deconstruction of his life. Is a private wedding at age 25 evidence of a secretive (and implied subversive) business future, as one writer suggested?

John Davidson Rockefeller was born on July 8, 1839, in Richford, New York, one of a group of late-1830s newborns who would mature together in the post-Civil War industrial boom to become the nation's first generation of magnates: Andrew Carnegie (born 1835), Jay Gould (1836) and J.P. Morgan (1837).

John D.'s mother, Eliza Davidson Rockefeller, was a devout Christian woman, raised in the religious fervor of the Second Great Awakening. Her strict abstinence from drinking, smoking, dancing, card playing and the theatre, inherited from her well-to-do temperance-movement parents, were passed unflinchingly to John D. He lived by those principles without compromise throughout his life.

John D.'s father, William "Big Bill" Rockefeller, was such an exact antithesis of the reliable and austere Eliza Davidson, it is strange that they married. William Rockefeller was a philandering vagabond, the product of a family of heavy-drinking hillbillies. He endlessly cloyed with his charm, good looks and loose dollars to endear those around him. Rumored to have left his girlfriend Nancy Brown to court and then marry the well-off Eliza Davidson, he brought the old girlfriend into the Rockefellers' lives as a "housekeeper" soon after the wedding.[1] In the first small house of his married life, Bill Rockefeller fathered five children—three by his wife and two by the housekeeper. John D. was born third in this progression, between two illegitimate sisters.

In spite of Rockefeller's frequent absence and an intervening move to Moravia, New York, the number of legitimate children eventually grew to six; the last, sickly from birth, died before her second birthday.

The Rockefeller family would move twice more—to Oswego, New York, and then to Strongville, Ohio, just southwest of Cleveland—as Big Bill's antics grew even more bizarre, and eventually estranged him from his family permanently. Eliza never divorced Bill, as this was not really an option in the mid-1800s. On June 12, 1855, William Rockefeller secretly married Margaret Allen in Nichols, New York, and started a secret bigamist life as Doctor Bill Levingston, a befitting name for a vagabond medicine salesman.

If much of John D. Rockefeller's success as the original American capitalist can be attributed to his good timing at birth, the complicated, contradictory and sometimes desperate conditions of his childhood may have been the necessitating element.

Rockefeller dropped out of high school in May 1855 at age 15. He apparently abandoned plans for college because his father was cutting back on support to his first family in preparation to marry Margaret Allen.[1] Rockefeller immediately took a three-month commercial course and went hunting for his first job, and at the same time, became the surrogate caretaker for his mother.

On Sept. 26, 1855, 16-year old John D. Rockefeller landed his first job, as assistant bookkeeper at the firm of Hewitt and Tuttle in Cleveland. Rockefeller reminisced late in his life, "As I began my life as a bookkeeper, I learned to have great respect for figures and facts, no matter how small they were … I had a passion for detail which afterward, I was forced to strive to modify."[3]

Unrelenting and unemotional in tracking claims, rectifying errors and collecting unpaid bills, the young Rockefeller had an immediate impact on the small firm of Hewitt and Tuttle. When Tuttle retired in early 1857, Rockefeller took over all of the responsibilities of the departed partner—at age 17.

On April 1, 1858, Rockefeller formed a produce-selling partnership with Maurice Clark, a 28-year old classmate from his brief commercial college days. Each put up $2,000 ($40,000 in modern values), and they hung out a shingle that read, "Clark and Rockefeller."

One year later, in a stinging move for Rockefeller, a silent partner was brought in for the sake of much-needed capital, and the firm was renamed Clark, Gardner and Company. This started a tense relationship that lasted 3 1/2 years before

Rockefeller expelled Gardner from the firm, and the name reverted to Clark and Rockefeller.

In 1863, Clark and Rockefeller were enticed to the emerging oil refining business by Sam Andrews, a personal friend of Clark and a fellow-Baptist churchgoer with Rockefeller. Clark and Rockefeller contributed $4,000 to the venture that was dubbed Andrews, Clark and Co. Envisioning the venture as a small side business, Rockefeller again "submitted" to being the unnamed partner. Yet, he immediately ploughed into this new business. A short time later, he ordered construction of the venture's first refinery, Excelsior Works, at a transportation crossroads that enjoyed access to Lake Erie and the soon-to-be-completed Atlantic and Great Western Railroad. Andrews became the refinery boss.

Some 20 refineries were operating in the Cleveland area by mid-1863. At that critical moment, and maybe for the last time in his storied life, Rockefeller did not enjoy any leverage over his competitors. But he would create his advantage quickly. If 19th-century philosopher Friedrich Nietzsche argued that the "language of good and evil" was not rooted in truth or reason, but in the "will to power," it was Rockefeller who embodied the philosophy. He started by imparting his extreme sense of thrift to the business and economizing, not from the office, but in the refinery works.

Sulfuric acid residues from the refining process were converted to fertilizer. The 42-gallon wooden oil barrels (eight of the nominal 50 gallons were lost en route to the refinery) were replaced by purpose-built containers. Piecemeal contracting was integrated through the hiring of staff and direct purchasing of materials. Within a year, refining had replaced produce as the most profitable part of the business.

Poised to plant his feet in the oil business and already shaping his style that eventually brought huge economic prospects, Rockefeller married Laura Spelman on Sept. 8, 1864. They would have a long and stable marriage, and their highly self-restrained and private home life was a stark contrast, possibly an oasis, from the vicious business climate at Standard Oil.

The quest for dominance and leaps of growth that drove Rockefeller soon began to tax Clark and his two brothers, so in February 1865, Rockefeller faced-down the Clarks with a classic take-or-pay proposition. Bids at the "auction" started at $500, quickly rose into the thousands, and then inched to $50,000 and beyond what Rockefeller thought that it was worth ... $60,000, $70,000 ...

The Clarks bid $72,000, and Rockefeller responded immediately with $72,500 ($750,000 in modern values).[1]

Thus, Rockefeller, at age 25, gained control of Cleveland's largest refinery.

Done with being "and Company," Rockefeller named the new enterprise Rockefeller & Andrews (see *The Life of Rockefeller*, pages 50-51). Showing undisguised bitterness toward his older partners, Rockefeller effused after the deal, "Then [the Clark brothers] woke up and saw for the first time that my mind had not been idle while they were talking so big and loud."[3]

Rockefeller thrived in a fractious atmosphere of dwindling oil feedstock and then oversupply; frequent refinery fires; improved rail transport and expanding markets; and competition that sprang up in six major locales—three inland (Cleveland, Pittsburgh and the Oil Regions of northwest Pennsylvania) and three on the seaboard (New York, Philadelphia and Baltimore). He soon expanded to three refineries and continued to integrate and optimize his operations. In 1866, he dispatched his brother William to New York to oversee exports from the refineries.

He never wavered from his premise that petroleum was and would be the basis of an enduring economic revolution.

Freewheeling with no mentors and no precedent to follow, and after getting rid of his lesser partners, Rockefeller deliberately took on a new partner in 1867, one who would prove very beneficial to Rockefeller, sharpening his plans, strengthening his resolve and broadening his capabilities. This lifelong cohort was Henry Morrison Flagler, who paid $100,000 for a one-third interest in what became known as Rockefeller, Andrews and Flagler. Ultimately, Flagler was instrumental in devising the most lasting of Rockefeller's strategies: railroad rebates.

Because his refining capacity was in Cleveland, further from the bulk of the consumers on the Eastern seaboard than any other major refinery, Rockefeller started at a clear disadvantage (see the color map at the center of the book).

The most naturally efficient refining site was Pittsburgh; crude oil was floated from the Oil Regions down the Allegheny River, refined, and transported in a straight shot to Philadelphia and New York by the Pennsylvania Railroad. Pittsburgh emerged as the main refining center. Crude refined in the Oil Regions and shipped as a finished product to the Eastern seaboard also enjoyed a

The Life of Rockefeller: 1839 to 1937

Age 19
...unnamed partner with Clark,Gardner and Co.

Mid-30s
...majority of Standard Oil empire in place

Date	Age	Historical Event
1839	—	Rockefeller is born
1855	16	Hewitt and Tuttle
1857	17	Hewitt (& Rockefeller)
1858	18	Clark and Rockefeller
1859	19	Clark, Gardner and Co.
1862	22	Clark and Rockefeller
1863	24	Andrews, Clark and Co.
1865	**25**	**Rockefeller & Andrews**
1867	**27**	**Rockefeller, Andrews and Flagler**
1870	**30**	**Standard Oil Co. (Ohio)**
1873	**35**	**Cleveland Massacre**
1875	**38**	**Monopoly nearly complete**
1882	**42**	**Standard Oil Trust**
1890s	50s	Rockefeller retires
1890		University of Chicago
1899	60s	Standard of New Jersey
1901		R. Institute for Medical Research
1902		Tarbell and muckrakers
1911	72	Supreme Court orders dissolution of Standard Oil Trust
1913	80s	Rockefeller Foundation
1937	97	Rockefeller dies

Mid-50s
...retires with net worth of $200 million

Mid-60s
...era of Ida Tarbell and the muckrakers

Age 85
...Rockefeller Foundation; net worth tops $1 billion

Photographs courtesy: Rockefeller Archive Center

Description	Concurrent Events
—	—
Assistant bookkeeper	California Gold Rush
Chief bookkeeper	Businessman's Revival of 1857
Produce business	—
Unnamed partner	Col. Drake well
Named partner	American Civil War (1861-1865)
Oil as "side business"	—
$72,500 "auction"	**Civil War ends**
$100,000 for 1/3 interest	**The Gilded Age**
10% of U.S. refining	**—**
25% of market	**Robert Nobel**
80% of market	**Russian oil**
90% of market	**Edison and electricity**
$200 million net worth	Marcus Samuel starts Shell
$75 million invested (ult.)	Royal Dutch
84% of market	Henry Ford's first car
—	D'Arcy and Persian oil
—	Spindletop gusher
70% of market	Wright Brothers' first flight
$300 million net worth	Anglo-Persian Oil
$1 billion net worth	World War I, Great Depression, FDR
"See you in Heaven"	Hitler, Saudi oil

built-in transportation advantage over oil refined in Cleveland, which had to travel the additional round-trip distance from the point of production.

Rockefeller overcame the transportation hurdle by exploiting the numerous transportation options afforded by the same location. First, he negotiated discounts from the Erie Railroad by shipping or threatening to ship his oil by barges on Lake Erie to Buffalo and then by train to New York.

The Pennsylvania Railroad gave Rockefeller a major opening by arrogantly overplaying its monopolistic hold on the natural refining center of Pittsburgh. While extracting high rates from Pittsburgh refiners (who had no other transport options) and even shipping raw crude all the way to refineries in Philadelphia, the Pennsylvania Railroad enjoyed the largest volumes of any railroad. In 1867, the Pennsylvania Railroad decreed that Cleveland would be "wiped out" as a refining center.

In a state of panic, Cleveland refiners began to move their refining operations east to the Oil Regions. The two northern railroads saw the obvious consequence of losing market share and also grew anxious.

At this critical juncture, Rockefeller stepped in and applied a master stroke that would tilt the competitive landscape in his favor for the next 40 years. He secretly negotiated a staggering 75-percent rebate for oil shipped by Cleveland refiners through the Erie Railroad system.

Rockefeller did not stop with this major concession. Next, he promised to provide 60 carloads of refined oil daily to the Lake Shore system; big, uniform shipments in massive quantities were good for the railroads, reducing both round-trip time to New York and the number of cars needed, and increasing gross revenue. Propped up by the Lake Shore deal, Cleveland soon surpassed Pittsburgh as the leading refining center.

Rockefeller lacked the refinery capacity to meet such an ambitious volume level, but the situation was soon rectified.

In 1868 and 1869, Rockefeller started a campaign to replace competition with what he called "cooperation." In his favorite characterization, "The Standard Oil Company was an angel of mercy, reaching down from the sky and saying: 'Get into the ark. Put in your old junk. We'll take all the risks.'"[6]

In 1870, the partnership of Rockefeller, Andrews and Flagler was replaced by a joint-stock company called Standard Oil (Ohio), with John D. Rockefeller as president. At this stage, Standard Oil controlled 10 percent of American oil refining.

In early 1872, rumors sprang up about a railroad cartel scheme being promulgated by the three most powerful railroads under the deceptive name, South Improvement Company. It targeted Rockefeller and Pittsburgh refiners. At that time, most Cleveland refiners were not making money because of another "oversupply," so Rockefeller swallowed up 22 of his 26 competitors in a 40-day period—the Cleveland Massacre. Standard Oil's market share was now 25 percent.

Next, Rockefeller went after Pittsburgh. In late 1874, he and Flagler met with the two most prominent refiners in Pittsburgh and Philadelphia. He believed that if the strongest refiners would yield, he could easily collect the rest. When the negotiations stagnated, Rockefeller offered the refiners the chance to come and review his books. What they saw was dumbfounding: Rockefeller could make a handsome profit at a price that was below their costs. Both capitulated, and in one day, Rockefeller got half of Pittsburgh's refining capacity and the leading Philadelphia refinery. During the next two years, he collected or shut down 20 of the 21 remaining Pittsburgh refiners.

By late 1877, Rockefeller had swallowed the remaining refiners in Philadelphia, Titusville and Baltimore, with only scattered holdouts in New York.

There was a great deal of empire- and wealth-building left to do—eventually he intimated his company into the railroads, started pipeline and oil producing businesses, and conquered foreign markets—but the legend of Rockefeller was already ordained.

At age 38, and only five years after the Cleveland Massacre, John D. Rockefeller controlled 90 percent of the oil refined in the United States, and by default, the world.

With the stone-faced portrayals common today in the popular press, few people realize that the triumphs of Rockefeller, for all practical purposes, were accomplished while he was still a youngster.

Even fewer realize that the Standard Oil Trust was not established until 1882, five years after his dominance was established. The company's influence probably peaked in the 1880s, when it controlled 84 percent of the U.S. petroleum retail market and produced one-third of its own oil (highest percentage ever). By then, however, the company was well into the bureaucratic stage, and the maze of Standard Oil companies proliferated.

In 1891, Rockefeller began turning the company over to his successor, John D. Archibold, and he retired in the mid-1890s, still in his mid-50s. The Sherman

Antitrust Act, which outlawed trusts and combinations in restraint of trade (targeting Standard Oil specifically), came and went with little consequence in 1890. By the end of the decade, Standard's stranglehold on world and U.S. markets began to erode slightly with the advent of California production; the Nobels and Russian oil; and Marcus Samuel and Shell Transport & Trading in Asia. The first automobiles were introduced in the same timeframe—an invention that would make John D. Rockefeller much wealthier in retirement than at work.

Rockefeller's only son and namesake, John D. Rockefeller Jr., joined Standard Oil after college in 1892. By then, the company had moved from Cleveland to New York City. John D. Rockefeller Jr. quickly gathered nominal titles as director and vice president, but he would never fill his father's shoes. In his words, "I was too squeamish." Ultimately, Junior opted for a full-time vocation as a philanthropist, devoting himself to the administration of his inherited wealth.[4]

Rockefeller himself spent far more time in retirement developing his philanthropic enterprise than he did in building the Standard Oil Company. By the time Rockefeller died in 1937 at age 97, he had donated more than $1 billion (in modern values) to found the University of Chicago. He commissioned the Rockefeller Institute for Medical Research (renamed Rockefeller University in 1965) that produced a string of medical breakthroughs and 16 Nobel Prize winners by the 1970s. His General Education Board ultimately dispensed more than $1 billion (in modern values). The Rockefeller Foundation, formed in 1913, today has an endowment of $3 billion and awards $100 million annually.[7]

Rockefeller was almost 10 years into retirement when Ida Tarbell used the emerging mass communication media to incite public sentiment against Rockefeller, inaugurating the so-called "muckraking" era. In a series of reports that seem to unravel in retrospect as a bitter personal attack (Tarbell's father languished in the struggling Oil Regions in the 1870s), Tarbell chronicled the rise of Rockefeller in 19 consecutive issues of McClure magazine, starting in November 1902. The focus of Tarbell's exposé and that of many Standard Oil detractors was Standard's endless participation in the gray business ethics of the day, namely railroad rebates and bribery of government officials.

As government caught up with the industrial boom and public exposures forced a change in its behavior, these practices graduated from gray to black, eventually becoming criminal offenses. During the same time frame, after Rockefeller's departure, John Archibold and Standard Oil applied bribes and rebate schemes with increasing aggressiveness. They contributed mightily to the feeding frenzy of the muckraking era.

Corporations then did not have publicity departments to handle this sort of attack. Rockefeller responded with silence, which in turn, implied his guilt, validated Tarbell's claims and invoked the public's rage.

Eventually, Rockefeller became somewhat of a master of public relations and, in the end, won considerable favor with the public. But much damage had already been done; the government had seized on the public sentiment, and the antitrust wheels were set irreversibly into motion. The public image of the oil business in the United States has never completely recovered.

Among the many industrial trusts of the day, most notably U.S. Steel, International Harvester (farm implements) and the Northern Securities Company (railroads), Standard Oil was ultimately singled out by President Theodore Roosevelt as emblematic of the dark side of trusts. The downfall of Standard Oil executives was that, drunk in their success, they treated the federal government as a meddlesome, inferior power.[1] Indeed, even while Roosevelt was making concessions and cooperating with other trusts, he was feverishly adding staff, collecting information and supporting progressive legislation directed squarely at bringing Standard Oil to heel. Like Tarbell, Roosevelt focused his attacks on Rockefeller personally.

On November 18, 1906, the federal government filed suit against Standard Oil under the previously toothless Sherman Antitrust Act (1890), naming as defendants Standard Oil, 65 subsidiary companies and a host of Standard executives, starting with John D. Rockefeller. They were charged with monopolizing the oil industry and restraining trade through railroad rebates, abuse of their pipeline monopoly, predatory pricing, industrial espionage, and secret ownership of ostensible competitors.[1] The suit called for dissolving the trust, breaking it up into its component companies.

The long, twisted path included hundreds of witnesses and thousands of pages of testimony; scores of antitrust suits from individual states; one massive fine from a Chicago judge (quickly overturned) that amounted to 30 percent of Standard's capitalized value; an initial judgment leveled by a St. Louis federal circuit court; and a subsequent appeal by Standard Oil. The final judgment was handed down on May 15, 1911, by Supreme Court Chief Justice Edward White.

Standard Oil was given six months to spin off its subsidiaries.

Ironically, by the time the verdict was handed down, Roosevelt had left office, Rockefeller was 20 years into retirement, and the court's decision was no longer needed. The monopoly was made possible only by virtue of the entire world's oil production being conveniently, and for an extended time, confined to one small part of Pennsylvania. With exploding production in Texas, Oklahoma, California, the Middle East, Southeast Asia and Russia, Standard Oil's stranglehold on the industry was not sustainable.

The breakup did have the enduring effect of strengthening the government's hand, providing a necessary balance in dealing with the massive U.S. industrial and capitalist machine that has lasted until today.

The breakup also had the unintended effect of ensuring Rockefeller's legacy. At the time of his antitrust defeat, Rockefeller was on par with Andrew Carnegie as the richest American and the richest man in the world.

In the decade following the breakup, Rockefeller's holdings increased in value by a factor of five, and his wealth surpassed that of Carnegie by a factor of two.

Modern Manifestations

The modern recombination of Exxon has conjured images of the old Standard Oil, but the comparison is simply not credible. Although many exaggerated figures have been reported (Table 1), the combined Exxon and Mobil holds less than 15 percent of U.S. petroleum retail sales. Their 4 percent of daily world oil production puts them behind several large national oil companies: Saudi Aramco, NIOC (Iran), Pemex (Mexico), PDVSA (Venezuela) and CNPC (China).[14] The combined oil reserves of Exxon and Mobil are even less, closer to 1 percent of the world total.

A fact, though, that would impress even Rockefeller, is that the emerging ExxonMobil has $200 billion in annual revenues—several times that of Saudi Arabia! Clearly, Exxon is still king when it comes to making cash out of even modest and, at times difficult, reserves and production. This is a reflection of enduring business savvy and seamless integration of oil and gas production, transportation, downstream marketing, products and technology.

The parallels between Standard Oil and Microsoft are more compelling and, again, point to a uniquely American phenomenon. Both companies controlled 80 percent of their markets. They both preferred to buy out competitors rather than to battle with them (by late 1997, Microsoft had made 60 acquisitions or alliances worth more than $2.5 billion).[15] In both cases, the bought-out companies and their employees did very well financially. Both companies presided over phenomenal drops in the price of their products: kerosene prices dropped from 88 cents to 5 cents during Rockefeller's reign; computers have dropped from $15,000 to $1,000 and even lower since 1985.

Table 1—ExxonMobil: A Resurrection of Rockefeller's Dominance?

Source	Est. ExxonMobil Share of U.S. Gasoline Outlets (%)
Mobil station owner in New Jersey[8]	35
U.S. News & World Report[9]	22
Houston Chronicle[10]	21
National Petroleum Network[11]	15
USA Today[12]	14
Exxon Mobil Corporation[10]	14
Energy Information Administration[13]	9

There are also superficial similarities such as the age of the industry and the dominance of the head honcho. Both leaders might be characterized as control freaks. Neither held a college degree. In Wendy Goldman Rohm, author of the recent "The Microsoft File,"[16] Bill Gates even has his equivalent of Ida Tarbell.

Matching Rockefeller's "angel of mercy" allegory, Gates contends that his company has ushered in the personal computer revolution and that its market success is the just reward for the service it has rendered the public.[17]

And many expect a similar result: Microsoft will be broken up, probably voluntarily. If, as in the case of Standard Oil, the sum of the parts is valued by the stock market as more than the whole (five times more, in the case of Standard Oil), then Microsoft stock, which sold for $100 per share in late 1999, may end up being worth $500 per share.

The Debate for the World

The role that petroleum plays in promoting personal and national wealth (see *Part I—Green*) puts it squarely at the center of great sociopolitical issues. When separating nations into "haves" and "have-nots," it is one of the main instruments of capitalism (or free enterprise, depending on one's point of view).

Standard Oil, from the emerging superpower country of the early 20th century, became both the symbol and the real standard-bearer of a political and social system. Ideologically, this was in clear contrast to the doctrines of fairness of other systems that were nurtured primarily among European intellectual elites, such as communism and socialism.

In many ways, among all countries of the world today, the U.S. is nothing but a bigger version of Standard Oil.

Although communism and similar systems have clearly failed in their applications, and though enterprise has clearly prevailed, the sentiments that carried the debate among these systems still linger today and may re-emerge.

It is not just the occasional and obscure conference of academic intellectuals and armchair revolutionaries expressing nostalgic revisionism for grand social debates. Many nations in Africa and South America are falling disproportionately behind; ex-communist nations, prime among which is Russia, find it very difficult to adjust; and, very important, almost all petroleum exporting countries have unraveled. Many exporting countries today have a mentality that is similar to that of the producers in the Oil Regions of Pennsylvania in the 1860s. Both seem to believe, as Rockefeller said, "the place where the oil was produced gave certain rights and privileges that persons seeking to engage in other localities had no right to presume to share."

As a result of mega-mergers in 1999, large petroleum companies have boiled down to only Exxon and two other very large post-colonial companies, Shell and BP. TotalFina Elf, a company put together in late 1999, became a quite distant No. 4. The next two largest companies, Chevron ($35 billion in revenues in 1999) and Texaco ($30 billion) would have to combine with each other and all of the next 10 non-national oil companies to be of the same magnitude.

In this narrowed set of players, Big Oil is and will continue to be at the center of not just the production and retailing of petroleum, but also the political events spawned by it.

The frustration felt by some politicians and the educated elites of petroleum producing countries is palpable. Woefully mismanaged and corrupt, these countries are helplessly watching the sell-off of a vital commodity, incapable of taking advantage of the revenues produced and unable to effect any significant control over the market.

Here is where the dark side of Big Oil becomes woven with its bright side. Technology, market savvy and management are Big Oil's forte and yet ...

What's good for Exxon—high oil prices—may not be good for the United States, the biggest oil consumer in the world.

What's good for the United States—low oil prices, which are the fuel for economic prosperity—is good for Exxon.

What's good for the petroleum producing countries—high oil prices—is good for Exxon (and every other oil company).

What's good for Exxon—low oil prices—may be disastrous for petroleum producing countries whose budgets depend entirely upon production revenues. (Exxon still profits from the high margin between production and retail.)

Exxon, Shell and BP have made and will always make money, boom or bust.

Part IV Red

War, colonialism and the access to oil

Street-to-street fighting in devastated Stalingrad, 1942. (*Courtesy: Sovietskaya Rossia, Moscow*)

In 1994, I was awarded an honorary doctorate by the Russian State Academy of Oil and Gas—the first time this honor was given to a foreigner. With two of my colleagues at Texas A&M University accompanying me as "witnesses" to "publicly state that I was worthy of this honor," I flew to Moscow.

Our hosts went out of their way to demonstrate just how much the award meant, and one of the prearranged activities was to visit with a man everybody referred to with undisguised awe: Nikolai Konstantinovich Baibakov.

Born in Baku in 1911, the 83-year-old Baibakov looked and definitely acted much younger than his age. His wit was unrelenting. His demeanor was imperiously cordial, exactly what could be expected from a former Soviet People's Commissar (minister) of the petroleum industry (1939-55) and former chairman of the giant Soviet oil and gas commission Gosplan (1965-85).

He was full of stories, and happy to tell them to this foreign "oil specialist."

In 1942, Stalin was told of what Hitler said: "We will get the oil from the North Caucasus or victory will escape us." On July 1, Stalin summoned Baibakov and minced no words: "You will go to the Caucasus and destroy the oil industry. If you leave for Hitler even one ton of oil, we shall shoot you. If Hitler does not enter the Caucasus, but in the meantime you have destroyed the oil industry, we shall also shoot you."

Baibakov told Stalin, "You leave me no choice," to which Stalin answered, "Think what to do."

Baibakov assembled a team and began methodically destroying wells, starting first with poor-producing wells and, while watching the German military advance, progressively getting to better wells. The best wells were blown up just before the German troops arrived at the giant reservoirs in Grozny and Maikop. Hitler's army, without fuel and under constant harassment by the Red Army, never went behind the main Caucasus. Baku was safe.

In 1944, Baibakov was again summoned by Stalin, who told him of his appointment as the People's Commissar of the petroleum industry. After Baibakov told Stalin he had no idea that he was a candidate, Stalin went straight to the point: "We know our folks. Tell me what is necessary to get a lot of oil for the defense of the USSR." Baibakov suggested massive investments in the Volga-Ural region. Immediately, Stalin called his man of action, KGB's notorious Beria. Shortly thereafter, the Soviet oil industry embarked on a massive drilling campaign in the area.

In 1963, Baibakov received the USSR's highest honor, the Lenin Award.

Baibakov recounts these and other stories in his book, "From Stalin to Yeltsin," published in 1998. – M.E.

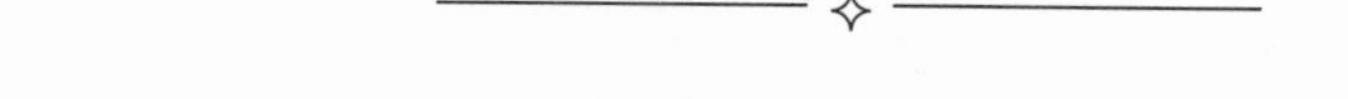

Red is the color of oil—as red as the blood of the millions who died in two great world wars and many other conflicts in this century. Central to the causes and prosecution of the wars was access to oil.

If there is one thing that characterizes the modern (previously known as industrial) era, it is the abyss that separates public myths, the nirvana in people's minds, from reality, which is full of conflict, competition and war. Ever since the European Enlightenment and the emergence of the bourgeoisie … ever since the quest for the rights of man found its highest manifestations in the French and American Revolutions, we have lived presumably in an egalitarian world where "all men are created equal" and "all are equal under the law."

These notions not only created and shaped expectations for modern societies, but also became the standard by which "civilized" behavior was measured. An indirect result may be the wide separation between public pronouncements in the media and what is really happening, so common today.

Observe two opposing lawyers in a lawsuit, the plaintiffs and defendants cordially exchanging banter and pleasantries among themselves during a trial break, only to pounce on each other five minutes later. Or, consider the even less credible claim in sports competition: "It's not who wins or loses; it's how you play the game." No one actually believes this.

Governments have mastered the art of public deception, found in its most unabashed admission by Josef Goebbels, Hitler's minister of propaganda: "A lie, often repeated, becomes reality." The trouble is that often those who make the public pronouncements, by all measures, also come to believe their own created reality.

It is almost certain that former U.S. President George Bush, guarding his "place in history," would still insist today—almost a decade later—that the Gulf War was for the "liberation" of Kuwait.

War, "father of all" according to Heraclitus, is the highest manifestation of conflict in pursuit of the competitive edge, the better life.

Natural Resources, Oil and War

There has always been a clear connection between natural resources and war. Although ideologies, religion and justifications on the basis of national and racial superiority have been offered as the high-road explanation for armed conflict, it is the search for natural resources and the coveting or defending of wealth that have most often precipitated war.

In the 20th century, oil has led to world domination which, in turn, brought about ideological and international supremacy. When oil was in shortage, it precipitated war, and the search for it guided the war aims.

The stage for oil was set for the last 25 years of the 19th century in the inextricable connection between economic development and industrial evolution. In the process, certain nations, especially those with colonial empires such as Britain and the resource-rich United States, benefited from rapid and disparate economic evolution and inequitable technological advances. Other European powers, Japan, and especially the rest of the world did not benefit to the same extent or did not benefit at all.

Russia was the first European country to feel the impact of oil development. While John D. Rockefeller and Standard Oil were planting the roots of the petroleum business in the United States—from production to refining to retailing, with the implications of the latter shaping social behavior—two of Europe's most prominent merchant families, the Rothschilds and the Nobels, won petroleum concessions in Imperial Russia.[1]

By 1880, the area around Baku on the Caspian Sea was producing most of the oil for Europe, and for the next 30 years, the Rothschilds and the Nobels would control Russian production. In 1895, Standard Oil and the two families came to an agreement that, if it held, would have united the entire world under one petroleum enterprise with allocated spheres of influence. The Russian government objected and the agreement was not implemented.[2]

This was one of the very few examples in which Russian assertiveness translated into a coherent management of Russia's national affairs. Chaos, capriciousness, a disdain for the Empire's non-Russian nationalities and contempt for Russia's own peasants characterized the rule of Czar Nicholas II. The entire country was sinking into desperation. Petroleum production and the "sellout" to foreign capitalists (the Rothschilds' Jewish heritage was always in the background) became the lightning rod for labor unrest and the underground proletariat movement.

One of the main labor agitators in Baku, Josef Djugashvili, wrote much later that his effort was to create "unlimited distrust for the oil industrialists." Both labor and ethnic unrest led to protracted violence in 1905, destroying the vast majority of Russian production and bringing into question the country's reliability for petroleum supplies.

Eventually, Djugashvili, who received his revolutionary credentials in the oil fields of the Caucasus, became better known as Josef Stalin.[2]

At the same time, the managing director of Royal Dutch, a Dutchman by the name of Henri Deterding, won over Marcus Samuel's Shell Trading Company in London, and Royal Dutch/Shell was born. In the preceding decade, Shell and Royal Dutch had discovered oil in the Dutch Indies on the islands of Sumatra and Borneo. Deterding, who can be considered the first creator of an integrated petroleum company, understood the connections between international petroleum resources, international markets, and the technology and management skills necessary for their success. Samuel was no match for Deterding and he gave up, bitter, but with essentially no fight.[3]

In 1911, the company, amid the Russian pre-revolutionary fervor, bought out the Nobels and the Rothschilds in Russia and became the petroleum power to be reckoned with on the European continent. The situation was therefore clear. With global interests from Russia to Asia and to the United States, Royal Dutch/Shell became a worthy adversary of Standard Oil.

The triumphs of Royal Dutch/Shell—a group masterminded by a Dutchman in Britain, and owned largely by Marcus Samuel, a British Jew—did not sit well with many Britons, especially Charles Greenway, managing director of Anglo-Persian. His oil company, under the control of Scottish merchants, had drilled in Persia (encompassing today's Iran and much of Iraq) and found evidence of oil in 1903. In 1909, a huge discovery was made. Repeatedly, Greenway

Winston Churchill—
Through two world wars, he worked relentlessly to secure petroleum for Britain's war needs. (Courtesy: Archive Photos)

invoked the national interest and spoke of Royal Dutch/Shell as a "foreign syndicate."[4]

The need for oil exploded at the turn of the century. A technological revolution, led by the automobile and the proliferation of the internal combustion engine, created a huge demand for new petroleum resources. Military needs, and a raging debate between advocates of a coal-fired versus petroleum-powered navy dominated the first decade of the century.

A young member of the British Parliament, named in 1911 as the First Lord of the Admiralty, Winston Churchill became convinced that a petroleum-powered navy would provide significantly greater flexibility, strength and speed than one powered by coal. In the acceptable words of the day he said, "The whole future of our race and Empire, the whole treasure accumulated during so many centuries of sacrifice and achievement, would perish and be swept utterly away if our naval supremacy were to be impaired."[5]

Central to this supremacy was petroleum.

The challenger was Germany. By 1911, the German Weltpolitik had, for 20 years, defined the country's national debate and its perceived role in the world for political and economic dominance. The British Empire was suffocating

German aspirations. In a fitting euphemism, Germany characterized its own ambitions as a movement for "world political freedom."

Conflict between Britain and Germany appeared imminent in 1911, but the precursor to future battles took place in Persia. Lord Gurzon, viceroy of India, had earlier called Persia a piece "on a chessboard upon which is being played out a game for the domination of the world."[2]

In 1913, Anglo-Persian asked for massive government aid to help it develop the Persian oil fields. Without such aid, Anglo-Persian would be absorbed by Royal Dutch/Shell. Greenway reminded everybody in Britain that the neutral Dutch government could easily fall under German influence. His company, on the other hand, "embracing as it did the entire oil fields of Persia ... should not pass under the control of a foreign syndicate." The government aid was given, and Britain was guaranteed petroleum supplies.[4]

As it turned out, the insinuations against Deterding, Samuel and their company were unfair. The two contributed mightily to the British war effort, including the clandestine overnight disassembling and transportation of an entire petrochemical plant from Holland to Britain. The plant was the only one that could produce petroleum-derived toluene, a necessary feedstock for TNT, which was in perilous shortage at the time.[3]

World War I

A war that everybody expected, but no one predicted to last as long as it did, broke out in 1914. It became the first worldwide conflagration, the Great War, known today and forever after as World War I.

It was supposed to last for a few months ... it lasted for more than four years. It cost the lives of 13 million people, destroyed all European economies and permanently changed the global dominance aspirations of colonial powers.[6]

World War I became a watershed event in human history; in the course of the war, the internal combustion engine, using petroleum exclusively, was pitted against horses and men. There was no contest.

Before the end of the war, armored vehicles provided a devastating punch, and aircraft brought in a new and formidable dimension.

The petroleum-driven war also destroyed a social upper class and replaced it with a lower class. In Russia, an entire social order was violently overthrown by the Bolshevik Revolution. But even in the rest of Europe, the destructive and powerful petroleum-driven, mechanized innovations erased the gold-tasseled, horse-mounted, battlefield-picture-perfect aristocratic dandy.

Britain, not endowed with oil of its own, secured access to and ownership of Anglo-Persian oil and maintained supplies of petroleum, for which the demand increased exponentially during the war. Battlefield vehicles, a unilateral conversion of the navy to petroleum power (Germany maintained a largely coal-fired navy), and especially the stunningly evolving war of air forces, tilted the balance of the war to Britain's side.

Beyond the military hardware, the use of petroleum spawned technological advances of all kinds. The pace of development was so frenetic that successive innovations quickly became obsolete.

When the petroleum-rich United States entered the war during the last 18 months of the conflict, it nailed the lid on Germany's coffin. Denied petroleum resources from outside the continent, and with the potential access to the recently developed Romanian oil fields thwarted by highly effective British military operations, Germany was choked.

The war was won, and the victory has since established the Anglo-Saxon culture—ideals and language, first benefiting Britain and then the United States—as the dominant influence in the world.

The conflict of 1914 to 1918 became World War I because it encompassed continents and many countries, but the controlling dimension was petroleum—where it was found, where it was transported, and the greatly escalating role that it played in the new technology of war.

The Period between the Great Wars

Germany and Japan, clearly willing to compete in the world of ideas, culture and economic supremacy, did not accept the newly established order after World War I. The two nations claimed ideological and even racial superiority, and thus, with mathematical certainty, moved toward World War II to overthrow the Anglo-Saxon victory.

If the importance of resources and oil brought about the Anglo-Saxon victory in World War I, it became clear between the wars that any nation with allusions to compete for ideological, cultural and economic supremacy would require access to economic resources, primary among which was oil.

The period was also marked by a substantial evolution of the petroleum industry and the addition of prolific new production areas.

First, Texas, Oklahoma and California became the petroleum producing states in the United States. Then, in 1911, the breakup of Standard Oil resulted in the creation of many petroleum companies that aggressively pursued exploration and production. Along with Royal Dutch/Shell, they literally spanned the globe in search of new resources.

Of course, no region was more important than the Arabian Peninsula. In 1932, Standard Oil of California (Socal) struck oil in Bahrain. Much of the Arabian Peninsula was then run by one Abdul Aziz bin Abdul Rahman bin Faisal al Saud, who after protracted tribal wars, consolidated his rule. Eventually, he became King Abdul Aziz and the father of Saudi Arabia.

Socal, using a fortune hunter named Harry St. John Bridger Philby as an intermediary (the same Philby whose son, Kim, would later become the most notorious Soviet spy in Britain), signed an agreement with Saud in 1933—with payment demanded in gold. Thus, Arabia entered the petroleum world, and the United States gained a commanding presence through American companies there.

In the upcoming World War II, petroleum supplies from the Arabian Peninsula and Persia fueled the Allied war effort in Europe and the Middle East. This would be a decisive factor in the outcome of the war, and a sharp contrast to the desperate situation that would eventually confront Germany and Japan.

World War II

World War II cost the lives of more than 46 million combatants and civilians in the period between the German invasion of Poland in September 1939 and the surrender of Japan in August 1945.[7] It was by far the bloodiest and most destructive conflict in the history of humankind.

The war exposed presumably civilized nations as capable of behaving in an unspeakably cruel, sadistically homicidal, insanely racist and ethnocentric manner.

To describe German and Japanese war aims, earlier historians have found morally convenient crutches in personifications of evil: Adolph Hitler and Emperor Hirohito. For how could the nations of Bach and Goethe, Kant and Lessing—and of refined tea ceremonies, calligraphy and poetry—perpetrate the Nazi and Japanese atrocities? They, themselves, must have been victimized and cajoled by deft, albeit evil, politicians. More recently, there have been more logical attempts to ascribe national responsibilities to particularly virulent and vicious events such as the Holocaust. This, in turn, also brought about echoes of Anglo-Saxon historical racism and anti-Semitism.

Yet, the evolution of German and Japanese expansionist ideologies have their roots in the growth of national consciousness in the 19th century. The constant comparison of limited national economic resources with those available to the colonial powers, pre-eminent among which was Britain, also encouraged expansionism.

Following the ascension of ***Hirohito*** *to the throne in 1926, expansion of the Japanese empire accelerated. By July 1941, Japan occupied almost all of Indochina, coveting the oil of Indonesia. The Japanese were increasingly hateful of America, Britain, China and the Dutch for depriving them of such a vital resource.*
(Courtesy: Archive Photos)

Unification of Germany, a country that had become highly fragmented under Prussia and Bismarck in 1871, was the first step. Emperor Wilhelm II advocated a campaign that would give Germany economic, political and military hegemony over Europe. If colonies in Africa and Asia were not available because they had been taken by other powers, then subjugation of all the nations situated between Germany and Russia and the final annihilation of Russia itself was the alternative.

Such ideas, fomented by exaggerated patriotism and racism, were the norm not only in Germany, but throughout Europe. The will of the countries' inhabitants to be conquered in the realization of big-power national aspirations was immaterial.

On June 22, 1941, German forces invaded the Soviet Union across a 930-mile front. ***Adolph Hitler's*** *general and specific aims were clear: "The sooner we smash Russia the better." The ultimate purpose of the invasion would be to advance "against the Baku oilfields." (Courtesy: Archive Photos)*

The Treaty of Versailles that ended World War I in 1918 confined German territory and ambitions to a minimum.

Hitler, whose ascendancy to power in 1933 has been examined by more modern historians than any other event, viewed the situation in this way: The German population was too small and the territory of Germany was too economically limited to guarantee the survival of his racially superior people in the international arena of racial competition. Immediate expansion of territory, and on a vast scale, was the only solution.[8]

Much later, in June 1941, two days before the invasion of Russia, Hitler was explicit: "What one does not have, but needs, one must conquer."[9]

In Asia, in the 1920s, Japan was experiencing a rapid economic growth. After the Sino-Japanese War of 1895 and the Japanese-Russian War of 1905, Japan gained *de facto* sovereignty over Manchuria while voicing similar sentiments expressed in Europe by Germany. Manchuria was labeled "Japan's life line" and was deemed a necessity to supply "living space" for the overpopulated Japanese islands.[10,11]

Following the ascension of Hirohito to the Imperial throne in 1926, expansion of the Japanese empire accelerated, first into China, and then throughout

Asia. At the time of the Pearl Harbor attack in December 1941, the Japanese-held territories covered an area as large as any in history. Following the ruthlessly aggressive takeover of Manchuria by Japanese forces between 1931 and 1933, an event known as the Manchurian Incident, the territory was renamed Manchukuo.

Heralding things to come, the level of atrocities perpetrated by the Japanese military on the populace was at par with the most murderous actions by Hitler's Special Task Forces in Eastern Europe and during the Russian invasion.

After the Manchurian Incident, there was no stopping the Japanese expansion. Much of China fell in 1937 after a protracted campaign (Peking was occupied then), and it remained a puppet state until 1944. By July 1941, Japan occupied almost all of Indochina.

Coveting the oil of Indonesia, and after the German occupation of the Netherlands in 1940, Japan immediately sent a delegation to Batavia, the capital of the East Indies. They were rebuffed by the local Dutch administration and by the American oil company, Stanvac, one of the main operators. (The Japanese would return to Indonesia in 1942, after open hostilities broke out, this time as conquerors.) This only increased Japan's self-image as being in a state of siege. The Japanese populace was increasingly hateful of the ABCD countries (America, Britain, China and the Dutch) for depriving Japan of such a vital resource.

President Franklin D. Roosevelt concluded that the only way out of the difficulties of the world was to cut off supplies to the aggressor nations "particularly ... to their supply of fuel to carry on the war."[2]

Formalizing its alliance with Germany and Italy, Japan signed the Tripartite (Axis) Pact in September 1940. This step was clearly aimed at the presumably neutral United States.

This move by Japan was not particularly well thought out, especially since the Japanese depended to a large extent on the United States for its oil supplies. Roosevelt, still hoping to avoid war, did not impose a complete oil embargo on Japan (in spite of considerable political pressure) until hostilities escalated in July and August 1941, leading to Pearl Harbor.

Cutting off Japan's oil would leave them with, at most, a three-year supply, already stockpiled.[7] Before the embargo, Japan had undergone a serious dilemma: whether to invade Russia from the East, taking advantage of German pressure from the West, or to keep moving southward. It opted for the latter, progressing toward Indonesia.[10]

On June 22, 1941, German forces invaded the Soviet Union across a 930-mile front. Hitler's general and specific aims were clearly expressed earlier, in September 1940: "The sooner we smash Russia, the better. The operation makes sense if we smash the State to its core with one blow. Mere conquest of land will not suffice." A total of 120 German divisions out of the available 180 formed a three-pronged attack, the first against Kiev in Ukraine and the second toward Moscow. Once the first two groups linked up, the third would advance "against the Baku oil fields."[7] (See the color map at the center of this book.)

U.S. President ***Franklin D. Roosevelt*** *came to the conclusion that the only way out of the difficulties of the world was to cut off supplies to the aggressor nations, "particularly... to their supply of fuel to carry on the war."* *(Courtesy: Archive Photos)*

The vicious German military operations, the killing machines of the SS troops and the local non-Russian Nazi paramilitaries, bent to eradicate all Jews and Bolsheviks, found their match in the incredible resistance of the Red Army and its ability to retrench, regroup and counterattack. Scorched-earth policy became the mainstay of Soviet strategy. The Russian populace was massively involved in resisting the invaders. A month into the campaign, observing the progress of the war, German Admiral Canaris (head of the secret code department) confided to a subordinate, "No one has ever succeeded in defeating and conquering Russia."[7]

A horrific military stalemate ensued. Leningrad, earlier and later called St. Petersburg, was encircled by the German army, and its people almost starved to death. In December 1941, just as Japan was launching its attack on Pearl Harbor, a Russian counteroffensive, aided by the relentless winter, repelled the German stranglehold on Moscow. German troops had come within a few miles of Moscow from three sides.

Facing a seemingly impenetrable front that spanned the Soviet Union from north to south, Hitler planned a summer offensive in the south expressly to

On July 1, 1942, ***Josef Stalin*** *summoned Nikolai Baibakov, eventually the Soviet minister of the petroleum industry, and told him: "You will go to the Caucausus and destroy the oil industry. If you leave for Hitler even one ton of oil, we shall shoot you."* *(Courtesy: Archive Photos)*

secure oil supplies. In July 1942, German forces pushed through the front, and in early August, they crossed the Kuban River. The threat to the Caucasus oil fields was now imminent. On August 9, the German army reached the giant oil field in Maikop at the north foot of the mountains.

There, they found that the Soviets had blown up all petroleum facilities and wells. The same day, they arrived in Krasnodar; the story was the same there and everywhere else. The Germans never reached Grozny, the center of petroleum production in the Caucasus. Baku would not be occupied.

Hitler Was Deprived of His Prize

Instead, the German army entered into a mortal and grinding struggle with the Red Army in the battle of Stalingrad. The ferocity of the combat was both primeval and decisive. Stalingrad held out in street-level, hand-to-hand fighting, and a counteroffensive by Soviet forces in the winter of 1942-43 encircled the German armies that were laying siege on the city. On Jan. 31, 1943, the Sixth German Army, which had fought the extraordinarily bloody battle of Stalingrad, surrendered.

For the Germans, the beginning of the end had begun.

Eventually, the Soviet army would push into Germany and all the way to Berlin. Although the atrocities against captives and the Jewish people in the conquered lands defy all presumed norms of human behavior, the years of war between Germany and the Soviet Union produced far and away the largest share of casualties, measured literally in tens of millions.

In Asia, the entire Japanese foray into Indonesia and Southeast Asia was for oil. That was the reason Japan had "gone to war."[2] By the time Japan gained complete control of the islands in March 1942, Dutch and American engineers who manned the petroleum production operations and refineries (Shell in Borneo and Stanvac in Sumatra) had blown up the facilities. But in the absence of a military counterforce, it did not take long for the Japanese to start producing petroleum again. Almost all the oil and refined products were shipped to Japan.

The distance that the oil had to be transported over the open seas proved decisive and not on the side of the Japanese. The Americans, reeling from the humiliation of Pearl Harbor, got the upper hand after the battle of Midway in June 1942. For the Japanese Imperial navy, it was an ignominious defeat. Four aircraft carriers were sunk, and the rhetoric of invincibility by Premier Tojo and others was markedly reduced.[7]

The American naval campaign took over from that point. The war, first of attrition and then growing in intensity, targeted oil tankers in particular.

By early 1944, the sinking of oil tankers outpaced their construction, and oil imports to Japan fell to less than 50 percent of the level from a year earlier. By early 1945, the imports stopped completely.[2]

The Japanese Imperial navy had been exceptionally handicapped and was no match for the American fleets, whose petroleum supplies were uninterrupted. Flying time for the Japanese air force was drastically reduced. With its domestic supply of petroleum almost non-existent by the spring of 1945, Japan was ripe for defeat.

The atomic bombs dropped in Hiroshima and Nagasaki may have precipitated the Japanese surrender, as most historians assert, but the shortage of petroleum had already forced Japan to its knees.

The Post-War Period

World War I was won by the Anglo-Saxon powers because they had anticipated the need for petroleum and secured the supplies that were necessary to successfully prosecute their campaigns. World War II was a reaction by Germany and Japan to the domination by the victors of the previous war. In their ideology of supremacy, Germany and Japan understood that petroleum resources were vital and believed that such resources, even when found in other countries, were theirs by the right of the mighty.

Of course, they failed, and the period since World War II has consolidated this century, the century of petroleum. To the victors, especially the United States, have gone most of the spoils.

Petroleum has been used as a weapon by the United States in winning the Cold War. As discussed in *Part I—Green*, the Reagan Administration's real and shadow policy in the 1980s of low oil prices caused the demise of the Soviet Union and the communist bloc. With oil being the Soviet Union's only credible source of foreign revenues (even today, the petroleum industry accounts for 60 percent of the entire Russian economy), the reduced oil prices plus the "Star Wars" arms race accomplished the task.

Access to petroleum and unhindered movement of the commodity are crucial elements to both the welfare of the United States and the European Union, and to the health of the international economy. This is particularly important in Europe and the United States, where current domestic petroleum production represents considerably less than 50 percent of consumption, a percentage that continues to decline.

A Remarkably Transparent Crisis

Iraq's invasion of Kuwait and the Gulf War are examples of the length to which the United States will go to safeguard a smooth petroleum economy.[12] Saddam Hussein, vilified as one of the rogue leaders of the era, invaded his tiny, but petroleum-rich neighbor in August 1990. Petroleum markets responded violently and immediately. The price of oil went up considerably.

Saddam's calculated risk in invading Kuwait was reasonable. The attention of the world and especially that of the United States was focused on Eastern Europe and the monumental changes taking place there. Yet, recognizing the danger from the Kuwait invasion, the United States put together a very diverse coalition of nations and fought Saddam's forces back into Iraq by early 1991.

U.S. President George Bush described the Gulf War as a "defining moment" with implications of a new (higher) world order, but the highly selective application of superior military force in a world of many similar conflicts clearly brought up the petroleum connection—the old and enduring world order. The message was unmistakable: Disturbing or threatening to disturb petroleum supplies and trading would not be tolerated.

The stewardship of petroleum, perhaps the most vital commodity in the world economy today, has brought about both power and responsibility for the United States. This is the century of petroleum, and by implication, this is the American Century, *Pax Americana*.

Part V

Primary Colors

Money, people and technology

J. Paul Getty reviewing a map with King Abdul Aziz of Saudi Arabia, 1954.
(Courtesy: Archive Photos)

By the time Mark "took a package" in early 1999, we had become good friends. For 17 years, Mark and his wife traipsed around the world in the employ of Schlumberger, with stops in the Philippines, Singapore, China, Pakistan, Mexico and Venezuela. As natives of North Dakota, they welcomed their first domestic assignment, a transfer to Houston, in early 1998. If nothing else, it would be a good thing for their three young daughters.

Being laid off was a twist they hadn't expected.

A highly competent, high-quality character in his early 40s, Mark was completely unfazed by the event (after all, the oil price had dropped to $10 a barrel), and he even welcomed the change of pace.

As it turned out, he decided on a whole new career—in the funeral business.

In just three months with a major funeral home chain, Mark moved from an entry-level sales position to mid-management. This is not surprising if you know Mark, but consider some of the details.

Early in his new job and just getting his feet on the ground, Mark pulled together and charted some local sales and growth figures in Excel. He presented them to his workgroup in PowerPoint, using a computer projector. On the basis of this single presentation, he was dubbed the resident "computer techie." Now, I know Mark pretty well, and frankly, I'd rate him at the lower end of the techie ladder.

The whole event struck Mark as weird, as it did me.

The thought flashed to my brain. I'd spent the past six years developing research programs at Texas A&M University and the University of Houston, all the while sounding the alarm that the oil industry is under-researched and not nearly as high-tech as we claim. How did I reconcile my claims with this story?

Then it occurred to me. The strength of the oil industry is not research or technology development per se; it is that we innovate—relentlessly—on the basis of technological advances, no matter where the advances come from. If Bill Gates can make a better computer, my thinking goes now, we in the oil business can probably put it to use faster and better than any other industry.

Mark relays another interesting story. It seems that the funeral home's tracking system had a small breakdown, resulting in one of their (deceased) clients being misplaced. This upset the (still alive) relatives tremendously.

Mark had only overheard the incident (the shouting of an angry relative, I think), but nonetheless, he took the initiative to raise the issue at the next staff meeting. His question was obvious and simple: What are we doing to ensure this does not happen again?

The response? Blank stares. No one had any intention of doing anything.

Mark implemented a small change in the standard work flow that did the trick.

For people in the oil industry, isolating and solving problems is so natural, like drinking water, that we don't even realize we are doing it. – R.O.

Money (lots of it), technology (basic, but demanding), and people (special ones) are the primary elements of the petroleum industry.

Baseball, somebody once said, is a simple game. You throw the ball, you catch the ball, and once in a while, you hit the ball. If you do these things well, you make money. The petroleum business has a lot in common with baseball. You find the oil, you drill for oil, and you produce the oil. If you do these things well, you make lots of money.

Except for one problem: Hitting a major league fastball is considered one of the most difficult feats in all of sports. Hitting home runs is plainly extraordinary. There are many analogies to the petroleum business here, too.

From the very beginning, the business attracted extraordinary characters—risk-takers, capitalists, colonialists, imperialists and just plain fortune-hunters, all in search of a "home run." In the process, they had to endure hardship, working inside incompatible and, at times, hostile cultures.

Especially today, with mature and aging production in North America and Europe, the oil business has often moved to countries that have little in common with the Western world. Algeria, Angola, Colombia, Iran, Iraq, Libya, Nigeria and others are major petroleum producing countries with intractable social and political conflicts. International petroleum professionals live and work there, helping to produce a commodity that is not only vital for the developed world, but is also the only credible source of income for those countries. Often, oil people are not merely inconvenienced and harassed. On occasion, they have been kidnapped, and in some cases, murdered.

If money, technology and people are the primary elements of this industry, why must countries like Saudi Arabia, a self-professed devout and highly exclusionary Moslem nation, and Venezuela, with a populist and leftist government, come to the Western infidels and capitalists for help?

They make money, but not enough.

They have technology, and they have spent lots of money on it. In fact, they have built entire technology infrastructures expressly to reduce their dependence on the United States and Western Europe. Still, they lack technology momentum and remain heavily dependent on the West.

They have people, many trained at the best universities in the United States and other Western countries, but more often than not, these people are missing essential skills.

What, then, makes the petroleum industry tick? What are its primary colors?

Money

The petroleum industry and petroleum producing countries have the money, or at least they have very large incomes. Even with the stinging price fluctuations of the past 15 years, the petroleum companies have been some of the world's most successful enterprises, and the countries have clearly earned many times the amount they would have without petroleum.

Yet, large cash flows require large investments. This is true in any industry, but particularly in the petroleum industry.

The magnitude of investment can be gauged with what we have called the activation or reactivation index. This is a measure of the total investment required to establish access to new oil, expressed in dollars per barrel per day of stabilized production.

For example, Iraq publicly announced in 1999 that it would seek $30 billion in outside investment to triple production from 2 million to 6 million barrels per day, meaning that its development costs were $7,500 per barrel of oil per day of new production. The analogous figure, the activation index, was $3,500 in Saudi Arabia. In almost all petroleum-exporting countries, although the reservoirs are prolific, local culture and deficient infrastructures result in high costs.

The activation index can also be very high in west Texas, on the order of $10,000. The costs are low, but the area is mature and the most prolific reservoirs have been depleted. Although a well might have been drilled for $500,000 in 1999—inexpensive by world standards—a good production rate would be 50 barrels of oil per day. Contrast this with Mexico, where the average well produces almost 600 barrels of oil per day, or Saudi Arabia, where the figure is 6,000 barrels of oil per day.

The activation index increases substantially if oil fields are aging or the local infrastructure is expensive and cumbersome. Thus, attracting investment money requires the presence of oil (reserves), prolific reservoirs capable of delivering large production rates, and low activation costs.

The world average activation index in 1999 was $2,000 per barrel of oil per day. Development costs on the continental shelf in the shallow Gulf of Mexico were the lowest in the world, at $1,000 per barrel of oil per day. In the deepwater Gulf of Mexico, by contrast, the activation costs were as much as $9,000 per barrel per day, but ongoing technology efforts promised to cut this figure by 30 to 50 percent during the subsequent five to seven years. This would tilt the investment balance toward the deep water, at the expense of Saudi Arabia and similar countries (Iran, Iraq, Libya and, to a lesser extent, Venezuela), the only other theaters with sufficient reserves to match the appetite of the new supermajor oil companies.

The pace of spending in the petroleum industry at the start of 2000 suggests that the total capital investment in worldwide exploration and production will exceed $1 trillion within the next 10 years.

With the world on an insatiable energy binge and with healthy cash markets, raising these investment dollars should be no problem. In practice, however, underwriting energy investments is not that simple, and there are some undercurrent problems.

"The Color of Oil" *– original art by Armando Izquierdo*

***"Green"** – original art by Armando Izquierdo*

"Black" *– original art by Armando Izquierdo*

"Red, White and Blue" *– original art by Armando Izquierdo*

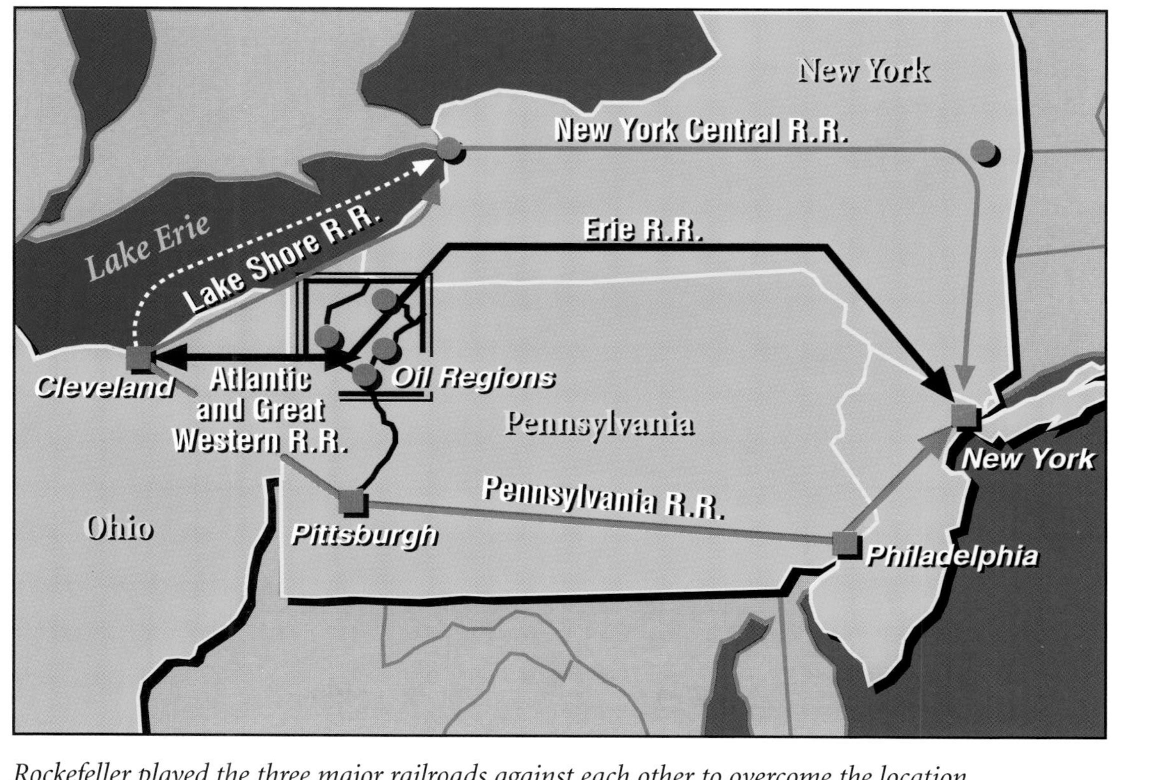

Rockefeller played the three major railroads against each other to overcome the location disadvantage of his Cleveland refineries.

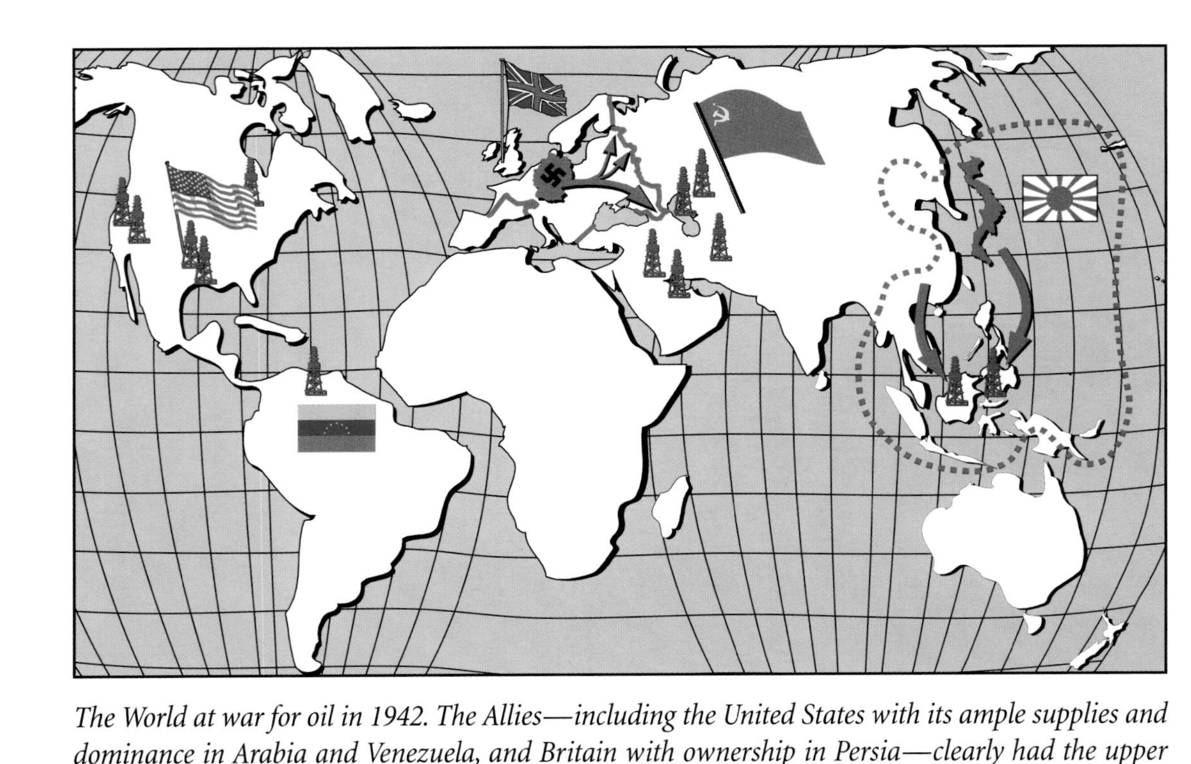

The World at war for oil in 1942. The Allies—including the United States with its ample supplies and dominance in Arabia and Venezuela, and Britain with ownership in Persia—clearly had the upper hand. Japan moved toward the Dutch Indies and had to prosecute a protracted and ultimately unsuccessful naval war. Germany invaded the Soviet Union, specifically for the Baku oil fields. The Soviets resisted mightily, the German campaign failed, and the outcome of the war was inevitable.

"Red" *– original art by Armando Izquierdo*

"Primary Colors" *– original art by Armando Izquierdo*

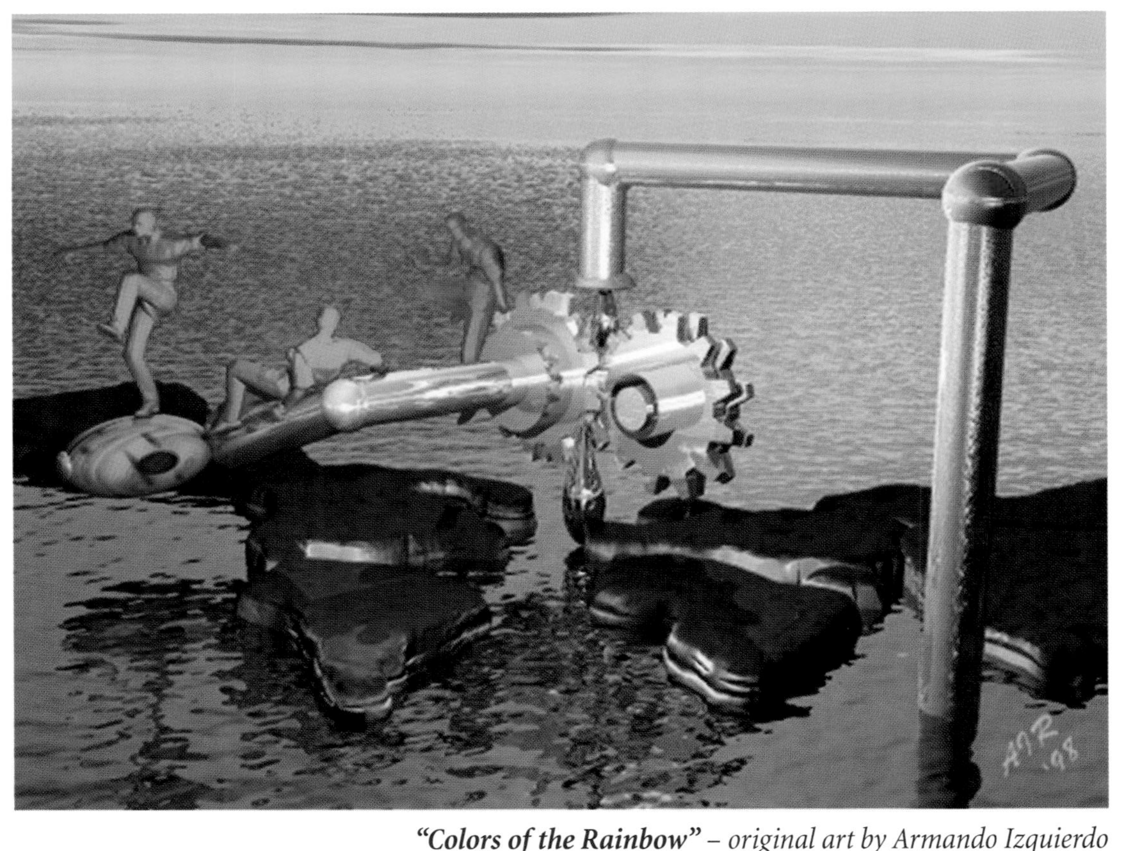

"Colors of the Rainbow" *– original art by Armando Izquierdo*

"Yellow" *– original art by Armando Izquierdo*

***"New Green"** – original art by Armando Izquierdo*

"Purple" *– original art by Armando Izquierdo*

In late 1999, when the physics of oil production depletion suggested that heavy investment in oil and gas development was needed to prevent future shortages, the investment climate was clouded by unjustified and destabilizing fluctuations in the oil price. For example, based on a couple of phantom events, oil prices dropped by a whopping $5 per barrel in two days in mid-October. The same frivolousness continued the subsequent week, but on the other side. Oil prices went up sharply for three consecutive days; the increase was attributed to a possible tropical storm in the Atlantic, which, in the worst case, would hurt production for only a few days, and the "approaching" cold-weather season.[1]

The decision to drop Chevron (along with Goodyear, Sears and Union Carbide) from the Dow Jones industrial average in November 1999 spotlighted the already finicky relationship between Wall Street and the petroleum industry. This gave the perception that Home Depot, which was added to the Dow Jones average on the same day, was more representative and more fundamental to the economic future of America than Chevron. This is clearly an aberration and a portent of misplaced emphasis couched in ignorance.

Nonetheless, large amounts of money have always flowed freely in the oil industry, with groups of investment banks routinely lending $300 million to $500 million and more for projects around the world. It is noteworthy that multinational companies, not the petroleum producing countries, are the prime recipients of these credits. What are the reasons?

A comparison of BP, Exxon or Shell to Saudi Arabia provides a clear distinction. First, each of the super-major producing companies has many times the cash flow of Saudi Arabia, the biggest of the biggest petroleum producers. This, in itself, is crucial, but the comparison is even more stark when the respective investment-raising capacities of multinationals and producing countries are considered.

The flow of money requires the "rule of law." How can it happen without proper assurances from the host government? Specifically, how can it happen if the host government itself has already been responsible for wholesale changes to the rules, empty populist sloganeering and, the most radical act of them all, nationalization?

This explains why BP, Exxon or Shell, legal entities that operate strictly under the rule of law, can borrow much more money and at much better rates than petroleum producing countries who, despite being sovereign states, have come to the brink of bankruptcy many times. The widely held notion that these countries can change the rules as they go is devastating. In the heady days of the late 1970s and early 1980s, when unfathomable petroleum cash flows were squandered, borrowing led to chronic problems and dealt a fatal or near-fatal blow to several overeager and lax international banking institutions.

Petroleum producing countries are also confronted with intense competition for investments and reinvestments in their economies. The leading competition, from Venezuela to Indonesia, from Libya to Iraq, is not from investments in external companies or even in other countries. The leading competition is from internal sources: corruption, mismanagement and "social spending."

Even Saudi Arabia must use a sizeable fraction of its petroleum income to placate a potentially restive citizenry. The country has a massive resource base, but fulfilling the popular notion of "flooding the market" would require literally $100 billion in internal investments. This investment level represents 1.5 times the Saudis' annual revenues of $70 billion, and 10 or more years of profits. Given the social programs and other internal pressures, the required time span extends to perhaps 50 years.

A $100 billion internal investment by the Saudis is clearly out of reach; it is not even a remote possibility.

The amount of money that the oil industry moves and invests is exceptionally large, but there is a thunderous, almost arrogant, subtlety about the situation. In this era of political correctness, companies are much more likely to tout their few hundred-thousand-dollar contributions to the opera and the theater, other civic causes, or the ubiquitous environmental and safety issues. "Safety Is Our Business" trumpets in big, bold letters from a petroleum tank farm east of Houston.

Occasionally, a maverick executive—rarely from the largest multinationals—may cause front-page news by announcing construction of a $200 million downtown skyscraper or may enter the flamboyant circles to compete for the naming rights of an athletic facility. At the same time, the $1 billion his or another company invests in a deepwater platform in the Gulf of Mexico, with a formidable economic multiplier that implies a several-fold impact on the local economy, warrants barely a mention on the business page. Few of the beneficiaries of this economic activity realize that several of these mega-projects are announced each year in Houston alone.

The $1.2 billion Hoover/Diana offshore development project shown in Figure 1 provides a compelling physical contrast to more visible downtown structures. The 12 mooring lines, which appear thread-like in the figure and spread out in a tent shape, are actually very large chains. Each chain is 1.3 miles long and is made of individual links that are 3 feet, end-to-end, and have a 6-inch solid steel cross section.

Fig. 1—Exxon's Hoover drilling and production platform for the Gulf of Mexico is nearly the size of the Empire State Building in New York City. (Courtesy: John Perez Graphics & Design, Richardson, Texas)

Certainly, oil is money, and money is a requisite for success in the petroleum business, but money alone is not enough.

Technology

It is no accident that Royal Dutch/Shell and Jersey Standard (Exxon), along with the Nobels in Imperial Russia, were the first purveyors of research and development (R&D) in the oil business before the main research centers shifted to Houston in the 1930s and 1940s (Table 1). Obviously, this group is well represented among today's survivors.

Table 1—Shell and Exxon Petroleum Research Activities

Year	Company	Location	Research Focus
1880	Nobel Laboratory	St. Petersburg	Kerosene/ Transport
1895	Royal Dutch/Shell	Delft, Holland	Product
1914	Royal Dutch/Shell	Amsterdam	Gasoline
1916	Royal Dutch/Shell	California	Chemicals
1919	Jersey Standard (Exxon)	New Jersey	Refining
1922	Standard Development Co.	New Jersey	Engineer/ Patents
1927	Esso Research Labs	Louisiana	Chemicals
1928	Shell Development Co.	California	Chemicals
1929	Humble Production Research	Houston	Geophysical
1947	Shell Bellaire Research Center	Houston	E&P Research
1958	Esso Hamburg Research	Germany +	Downstream
1964	Exxon Production Research	Houston	E&P
1975	Shell Westhollow Research	Houston	Downstream

Technology is a much used and often misunderstood word in all industries. It is particularly intriguing in the petroleum industry.

If "protecting the environment" and "safety" are the buzzwords to appease the politically correct crowd, and if money and profit are the essential elements for the bottom-liners (who will always win the day), then technology and claims of technological prowess provide the ideological glue, and at times serve as a surrogate for religion within corporations.

In many industries, there is a clear correlation between research expenditures and the welfare of a company. At times, the products of R&D hold such value that a company may relinquish all current product lines in their favor. Yet, technology is neither a tangible possession nor a static state.

In the petroleum industry, from the early 1970s to the early 1990s, more than 80 percent of R&D spending in exploration and production was done by just 11 operating companies.[2] In 1999, following the large mergers, this group shrank to eight.

Many unique features distinguish the technology of the petroleum industry. First, there is little doubt that technology is crucial, and that deployment and integration of technology is essential to the industry's success. Yet, this technology is highly diversified and applied to industry segments with different needs. The scope is wide. Seismic exploration and processing, enhanced oil recovery and the construction of deepwater production facilities have little in common.

Why is it, then, that the petroleum industry, so technologically dependent, is the industry with the smallest R&D spending? The healthcare sector leads all industries, with 11 percent of sales going to R&D; the electrical and electronics industry spends 5.5 percent, and the chemical industry spends 4.1 percent. In this light, the petroleum industry's R&D spending of less than 0.5 percent of sales is striking.[3]

Until the early 1970s, the picture was different. Research budgets among petroleum concerns were comparable to those of other industries. Much of the research was directed at understanding the physics of oil production and, especially, devising mechanisms for improved or enhanced recovery. However, while upstream revenues increased substantially following the dramatic oil-price jump of October and November 1973, upstream R&D expenditures remained largely constant for almost two decades. In the early 1990s, they were downsized dramatically, to the point that operating companies today do little, if any, long-term critical research—even if they list R&D expenditures in their annual reports. A close examination shows that work labeled as R&D is fragmented and highly dissected, low-impact applied research or technology services.[2]

In the past, we have used a visualization of cumulative technological developments, introduced by R.N. Foster and widely known as the S-curve, to explain what appears to be a contradictory situation: an industry heavily dependent on technology, yet with little R&D activity.[4,5]

Consider the construction in Figure 2, in which cumulative technology developments are plotted as a function of time. Technological needs are shown as dotted lines labeled "U.S. Fields" and "World." These lines can represent the technological needs of a company, an industry or society itself.

How does the S-curve (shown as a heavy black line) evolve? At Point 1, companies or industries recognize a need and start gearing up, recruiting people, building laboratories and so on. Soon, a critical mass of activities is reached, and developments pick up pace, often outstripping the demands (Point 2). At some point, the good ideas, enthusiasm and budgets for the research begin to wane—the technology matures—and the pace of development tapers dramatically, resulting in a second flat portion and completing the S-curve (Point 3).

At this point, a properly managed R&D organization recognizes the maturing process and gears up for the next S-curve. Companies in all industries live and die by this sequence of successive S-curves.

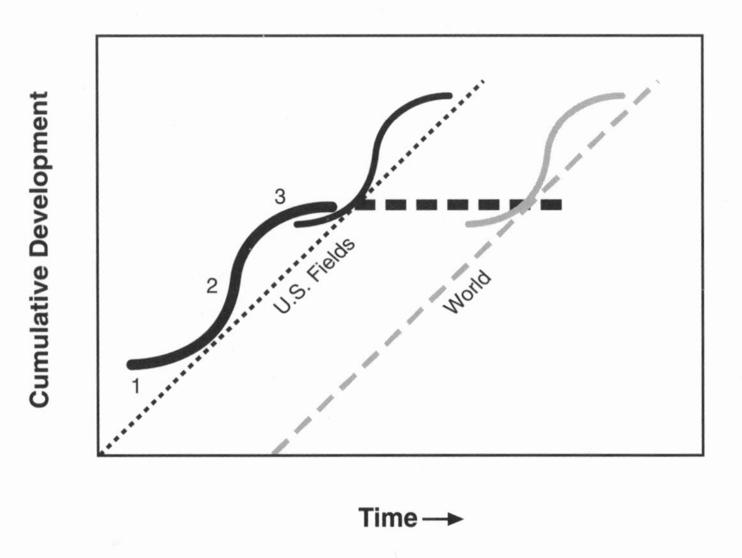

Fig. 2—Technology S-curves explain the disparate roles that technology can play in the petroleum industry.

The petroleum industry has had a substantial reprieve from the demands of this sequence—one that has shaped the past 25 years and will dominate the foreseeable future.

The entire body of petroleum technology can be described with S-curves, but with an additional slant. The 1973 oil-price trauma and subsequent opportunities resulted in a series of technology demand curves, one for each country or even different provinces within a country. The majority of these curves fall at one of two extremes shown in Figure 2: the technology demand is characteristic of either very mature fields ("U.S. Fields") or of new petroleum producing provinces ("World"). Few places fall in the middle.

To meet the needs of a less-depleted World, the petroleum industry can coast on existing technologies (the heavy dashed line in Figure 2), whose S-curve has already crossed and fallen below the technology demand curve for U.S. Fields. New technologies necessary for the mature fields are not cost-competitive in the prevailing market.

When will the flat portion of the S-curve cross the World technology demand curve? That is the big question.

For reasons ranging from pragmatic to patriotic, national oil companies have long implied that technology is vital. Venezuela's Intevep was set up immediately after the industry's nationalization in 1974 "to provide technology for the national petroleum and petrochemical industries, in order to reduce to a reasonable level their technological dependency on foreign resources."[6] Such organizations have not made much of an impact in the international arena, or even internally. Competition among multinational oil companies peaked in the 1970s and 1980s, and the oil-field service sector was highly fragmented. This gave the newly nationalized companies a sense of assertiveness as they played one company against another. The consolidation of the 1990s—which swallowed up even major long-term players such as Amoco, Arco, Mobil and Dresser-Sperry Sun, and internationally, Elf and Fina—made it clear that the national oil companies still depend on foreign companies for much of their technology.

What should not create any confusion, but unfortunately does, from soccer moms all the way to the U.S. president, is the role technology can and cannot play in extending domestic petroleum reserves and, especially, production. These concepts are central to our energy self-sufficiency, economy and national security.

Figure 3 shows the history of oil production in the United States from 1949 to 1997.[7] The lower (gray) line, depicting production from the Lower 48 states, follows a classic bell shape. Annual oil production peaked in 1970 and has been declining ever since. Important technological innovations—including 3-D seismic data processing since the early 1980s and horizontal wells since the early 1990s—could not stem the decline.

However, the debut of the North Slope of Alaska as a petroleum producing region had a great impact. The top curve in Figure 3, representing annual oil production for the entire United States, shows the impact of Alaskan production. The cumulative incremental production from Alaska (the area between the curves) from 1977 to today is 13 billion barrels of oil. This equates to $250 billion in direct revenues to the petroleum industry and the country.

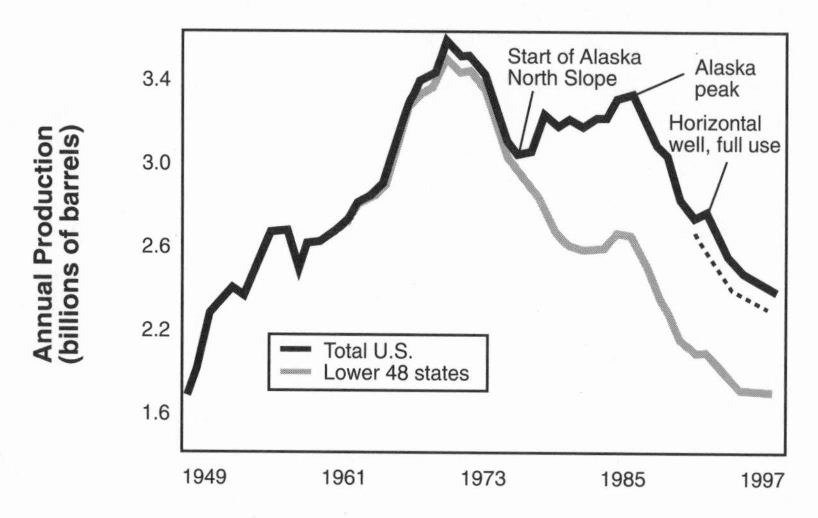

Fig. 3—U.S. and Lower 48 annual production shows clearly that new reservoirs are the name of the game.

Interestingly, the impact of a new technology such as horizontal wells can also be tracked at the national scale. The "bump" in Figure 3 after 1990 represents $25 billion in incremental revenue, which is respectable, but only one-tenth of Alaska's influence. More to the point, even a seminal and robust technology like horizontal drilling could not come close to stemming the national production decline.

Another important element to the story is told by a comparison of the oil reserves. From 1977 to 1997, oil reserve additions in the Lower 48 states totaled 36 billion barrels, whereas in Alaska, only 8.8 billion barrels were added. Clearly, the ability to book reserves does not correlate well with the capacity to produce oil.

Finding new reservoirs is the name of the game. Technology that provides access to them is dominant over any exploitation technologies for mature fields, no matter how attractive and adroit the latter can be. Accessing the fields of Saudi Arabia—and in the same breath, Iraq, Kuwait and the former southern Soviet Republics—is one option. The patchwork of kingdoms and tribal jurisdictions, difficult infrastructures and distances to market often make this option tenuous and sometimes unthinkable. The second option—one that beckons very seductively—is deep water, first in the Gulf of Mexico, but also in Brazil and West Africa.

The U.S. Geological Survey estimates that the deep offshore Gulf of Mexico holds 46 billion barrels of "technically recoverable" oil reserves.[8] Taking a cue from the Alaskan experience, these yet-to-be discovered reservoirs could propel the U.S. production to levels not seen before.

The super-majors are not the only companies elephant-hunting in the Gulf of Mexico. (An "elephant" is nominally defined as a reservoir that holds more than 1 billion barrels of oil reserves.) Smaller multinationals such as Conoco and Phillips, and large independents such as Anadarko Petroleum, have set out in an admirable albeit quixotic search for this new motherlode of petroleum resources.

The deep water is a veritable trillion-dollar opportunity for the United States and its petroleum industry.[9]

People

In the last half of the 20th century, the world focused so much on technology that it became a given in the minds of many that technology equals value. This is certainly true, but it is not the entire story. Although technology is readily available and can be purchased (or even stolen in certain celebrated cases), it is technology momentum—the ability to innovate across a broad spectrum of technological progress—that creates value.

This implies that people with certain unique qualities are needed.

The disparity in the technology momentum of various countries—so prominent in the petroleum industry—is not an educational issue. The bulk of key positions in the Kazakhstan government, for example, are filled with Kazakhs who hold MBAs from prestigious Western universities. And just about every senior technocrat in national oil companies is an engineering graduate of a university in the United States. Yet, this does not adequately fulfill the people requirement.

National oil companies without clear-cut economic goals, such as profit, and with political ties to regimes with populist and ideological agendas, are the graveyards of would-be entrepreneurs and potential technological innovators. There are no incentives for anything else. Conversely, people from the same countries who are trained and acculturated by major transnational corporations do considerably better. No company is as successful in this regard as Schlumberger. The French/American company hires hundreds of new people each year, about two-thirds of them from some 80 countries outside the United States and Europe. After undergoing formal training for two to three years, these workers become quintessentially international.[10] Their rise in the Schlumberger ranks is completely unrelated to their national origin, resulting in an upper management that is astonishingly diverse.

The people issue has a major cultural dimension that is addressed in *Part VI—Colors of the Rainbow.*

Cultural propensity for technology momentum is not as simple as the separation between developed and developing nations. It also cannot be attributed solely to the recent colonial and imperialist history (a frequent cop-out in developing countries), nor can it be lowered to crass racial and ethnic stereotyping (gone frighteningly astray in the first half of this century and still occasionally heard in developed countries).

Rooted in complex historical, religious and even geographical factors, culture is everything. Even within presumably homogeneous regions, cultural differences are pervasive.

Everybody knows the joke, and everybody in Europe winks in approval every time they hear it:

> "The European Continent has all the makings of 'paradise' on Earth: the policemen are British, the car mechanics are German, the cooks are French, the Greeks are the lovers, and the Swiss are the managers. Stir the mix, however, and it is 'hell' on Earth: the policemen are German, the car mechanics are French, the cooks are British, the lovers are Swiss, and the Greeks are the managers."

Just about everybody in France, as well as many intellectuals throughout the world, consider U.S. management schooling and pop culture (from CNN to MTV) intrusive and destructive to local culture. They are openly offended that U.S. thinking dominates the world of business and even everyday life.

Others, however, consider the American way of seeing and doing things, and especially the demands for rule of law (even if it has to be fabricated locally from scratch) to be a gift to the world.

Murray Weidenbaum of Washington University said it eloquently: "American culture, warts and all, is the pacesetter for a great portion of the world's population."[11] The same axiom is pervasively applied to the relation between the United States and the global petroleum business.

Primary Colors

Understanding the primary colors of petroleum (money, technology and people) explains a number of things:

- It explains why the multinational oil companies have abandoned production in the United States in search of newer reservoirs elsewhere.
- It explains why the current industry consolidation soon will expose the weaknesses of national oil companies.
- It explains why Saudi Arabia and all others must reopen their upstream activities after all these years of nationalization, even though they may not want to.
- It explains the true role of technology, so important and central, but so often reduced today to a buzzword.
- It explains the role of people, and the qualities required for their profitable and effective participation in the industry.
- Finally, it suggests an obvious path for the oft-confused U.S. policy on energy, and especially, on petroleum.

Part VI
Colors of the Rainbow

The dominance of culture in the world of oil

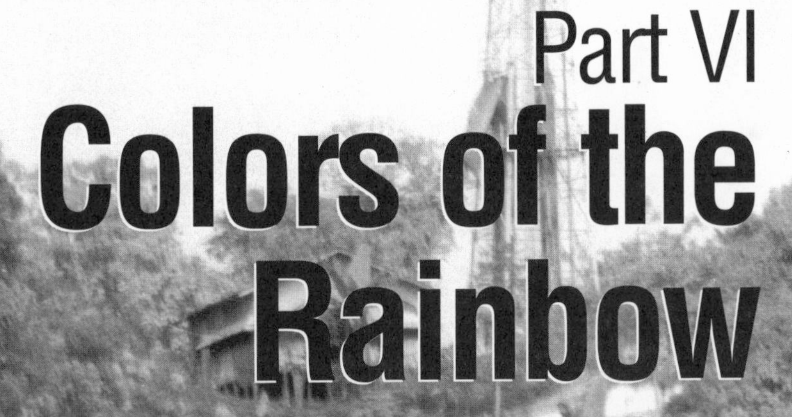

Drilling in Papua New Guinea.
(Courtesy: E&P magazine)

For years, first while I was with a major service company and subsequently as a senior academic, I gave short courses regularly in a major petroleum producing country. The training budget was beyond belief, perhaps more than that of the three largest multinationals, combined. The effectiveness was another matter, and the technology absorption was constantly in question.

My courses, covering three different subjects in my area of technical expertise, were well received, judging from the participants' reviews and the almost constant communication I had with the training director of the company and his staff. Several hundred engineers and technologists took my courses; some took all three courses at different sessions.

I became widely known in the company, and whenever its people needed information or a technical host for a visit to the United States, they would turn to me. I was on a first-name basis with several high-ranking executives who called often, sometimes just to chat. The entire activity amounted to very lucrative consulting for me, and the relationship seemed to be flourishing even more. I was becoming their "technical eyes and ears in the outside world" as I was told all too often. They made it clear that my prominence in the international petroleum industry added to my appeal.

One fateful day, the training director dropped a seemingly innocent suggestion. Why don't I come to the country as a resident "senior adviser on production technology," taking leave from the university? The assignment, supposedly to serve the entire company (not just the training arm), had the explicit blessing of the executive vice president and was positively alluring. With a generous contract in both money (perhaps the equivalent to the combined salaries of 20 local engineers) and perks, and a veritable palace for a house, I settled in.

It didn't take long for me to sense a difference.

Neither my host director nor any of the other executives who talked endlessly to me by long-distance telephone now seemed to have time for me. After several weeks of essentially doing little, I was given the assignment of writing an evaluation of their training curriculum and courses taught by international instructors. The idea of providing advice at the corporate level seemed to be fading away rapidly. None of my previous contacts would return my calls nor heed my direct and indirect requests for meetings.

An expatriate was succinct: "Your real worth was your presence in the United States. Your stature diminished immeasurably the moment you arrived here. How can you be good if you accepted to be here?"

Several months later, with my original host director already reassigned, it was apparent that my report on the training curriculum and, by extension, my evaluation of the company's technology proficiency and technology management, was largely ignored. For the few who read my report, its contents flew like a lead balloon. It was discounted without discussion.

Eventually, after I had left the country, the gossip was that I did not understand the company and the culture. My old relationship with them was never the same. – M.E.

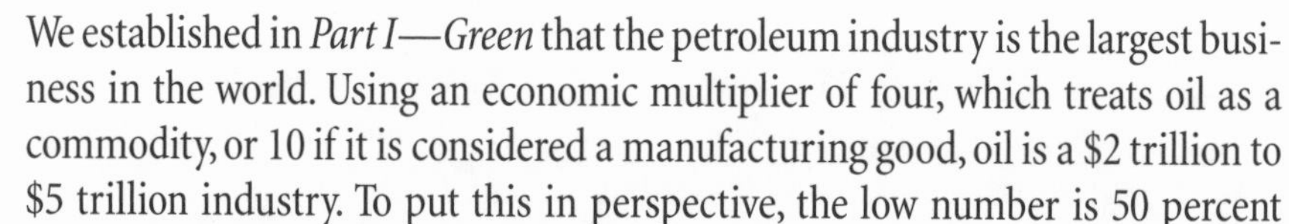

We established in *Part I—Green* that the petroleum industry is the largest business in the world. Using an economic multiplier of four, which treats oil as a commodity, or 10 if it is considered a manufacturing good, oil is a $2 trillion to $5 trillion industry. To put this in perspective, the low number is 50 percent larger than the size of the U.S. government budget; the higher number is comparable to the entire U.S. economy.

The players—countries and companies—form a huge rainbow of cultures, on both national and corporate levels.

They fall under three general categories: producers, consumers and purveyors of technology. In the past, players overlapped categories, and in the era of "Big Oil," a few companies had their fingers in everything. But lately, there has been an unprecedented polarization among producing countries, large multinational corporations and the companies holding the technology assets.

Historical Basis

Until the late 1950s, the industry, always colorful and always at the center of major geopolitical events, was relatively simple to understand. Half the cumulative oil in the entire history of the business had been produced and consumed in the United States. There were several big multinationals, mainly offspring from the breakup of Standard Oil, and a few others with colonial roots, such as Royal Dutch in Indonesia; Shell from the commercial prowess of the British Empire; Anglo-Persian (eventually, British Petroleum) in Iran and Iraq; and the predecessors of Total and Elf in French Africa. These multinationals held sway over much of the petroleum business and maintained a virtual monopoly over its technology.

The technology at the time was robust, but relatively primitive by today's standards: bulky drilling, phenomenological geologic observations, and a few subsurface measurements including electrical logging.

Investments for petroleum production outside the United States were always large, often requiring the building of not just oil-field facilities but complete infrastructures—entire towns for the employees and their upkeep in countries with the most primitive conditions. The industry was dominated by "alpha males" who worked and lived with the precision and discipline of military operations and tolerated the associated hardship.

The 1960s were a relatively quiet decade for the petroleum industry, but the time was marked by a rapid increase in worldwide oil consumption and the establishment of a clear relationship between per capita income and per capita oil consumption. To this day, one can position all nations of the world on an essentially straight-line plot of these two variables.

Also, emerging and post-colonial nations began to recognize the potentially enormous value of their petroleum resources. National oil companies were created and eventually generated large (not necessarily well-managed) amounts of income. They also became welfare agencies with massive and highly inefficient employment payrolls. Politically, they were cast in the role of guarding the national petroleum trust, and many were restricted by constitutional provisions, becoming almost as impregnable (in the most extreme case) as Mexico's Pemex.

Nationalizing the holdings and operations of the giant multinationals became a litmus of political awakening and nation-building in the 1970s. Eventually, this led to the formation of OPEC and, for a while, a perceived shift of wealth and power that alarmed the developed nations. Although they were theoretically allied to the West, Saudi Aramco and companies in assorted Arab sheikdoms, Iran's NIOC (before the collapse of the Shah), and Venezuela's PDVSA often asserted their power and virtually gloated that the former colonial and imperialistic nations were finally receiving their comeuppance.

Even these presumably friendly nations would occasionally come together to further a political motive, such as the Arab Oil Embargo to effect western support for Israel. That action (or merely the threat thereof) caused the first energy crisis of the 1970s and 1980s and became a watershed event. Not only did it show how panic for such a vital resource can cause considerable crises, it also established that even a very small over- or under-supply (in itself, often ill-defined) can cause price fluctuations of 50 percent or more, wreaking havoc on oil markets and rendering entire national budgets practically useless.

Oil, in the hands of nations that were hostile to the West, such as Libya, post-Shah Iran and Iraq, was supposed to be a weapon, but never really materialized as such.

It was blunted not only because concerted action was never really a possibility among such conflicting nations, but also because the energy crisis of 1973 actually helped the developed world, forcing the multinational oil companies to look for sources of energy outside the traditional petroleum provinces. The price of oil skyrocketed to more than $40 per barrel in the early 1980s, prompting major oil companies to actually publish predictions that oil prices would exceed $100 per barrel within a decade. High oil prices made many marginal fields attractive, and huge exploration budgets and development costs in difficult locations became quite bearable. The movement also became the impetus for a technological resurgence, major diversification of exploration and production portfolios, a galloping offshore activity, the creation of European oil powers such as Norway and Scotland, and commercial production in dozens of countries on all continents.

The multinationals took nationalization squarely in the jaw and kept forging ahead. With minimum protest and even less intervention by their governments, the multinationals established themselves as the key to international transaction, transportation and refining, and augmented their already imposing presence in the retail area. The substantially higher oil prices of the late 1970s and early 1980s allowed large investments in frontier areas. Cash flows were so great that several multinationals also ventured into business areas completely outside of their core specialties, although this was short-lived.

One of the most distinguishing characteristics of the emerging petroleum business was that it diversified and ventured outside the prolific areas of the Middle East and Venezuela. Countries such as Angola and Peru and unfamiliar countries with nightmarish infrastructures such as Papua New Guinea became scenes of exploration and, eventually, economically attractive production. Difficult areas such as the North Slope of Alaska, the North Sea and the southern tip of Argentina were no longer considered frontier petroleum provinces. So successful was this expansion to new places that practically every political as well as military conflict, in the public mind—rightly or wrongly—gained a petroleum dimension. Wasn't the otherwise inexplicable conflict between Great Britain and Argentina over the Falklands/Malvinas all about oil?

All the while, the oil fields in the continental United States were maturing. Although economy of scale, operational efficiency and an excellent infrastructure and work force still allow the United States to maintain sizeable production, this production is clearly tenuous and precarious. A good well in Texas may produce 20 barrels of oil per day and 500 barrels of water per day, while a typical well in Algeria and Iraq may produce 2,000 to 4,000 barrels of oil per day and virtually no water.

From the mid-1980s to the present, real supply and demand largely matched each other. However, oil prices have fluctuated wildly from as low as $10 per barrel to $30 per barrel. An announcement of a winter month in the eastern part of the United States that will be 2 degrees warmer than usual; or a mere 1 percent reduction in Asian oil consumption because of a much more profound economic crisis; or a pittance of embargo-allowed Iraqi oil sales; or just the *threat* of a U.S. bombing may cause a $5 drop or increase in the price of oil, virtually overnight.

The United States accounts for 25 percent of all oil consumption in the world. Its per capita consumption is almost 30 times that of China. Imports have increased steadily in the past 15 years, despite conservation and increased efficiencies. Although importing 50 percent of consumption was considered alarming by politicians and economists in the 1980s, imports actually reached this level without much fanfare several years ago.

The identities of the major producing and exporting countries haven't changed much during the past 20 years. Big producers with relatively small populations, such as Saudi Arabia, the United Arab Emirates and Libya, and countries with large populations (presumably "absorbers" of the petroleum revenues), such as Algeria, Indonesia, Iran, Iraq, Mexico, Nigeria and Venezuela, still dominate the movement of oil.

A striking correlation is the economic turmoil, often accompanied by political turmoil, in almost all of the latter countries. One thing is certain: Economic development and use of oil revenues for economic diversification have not materialized. If anything, the aphorism that warns against single-commodity economies appears to have found a textbook application in these countries.

Slowly but surely, these countries are reopening their petroleum prospects to the previously expelled multinationals, and to assorted new players. Understanding how the infusion of new capital and especially new technologies can boost a country's petroleum activities is not difficult. Yet, these activities frequently result in internal political repercussions.

The 1990s also witnessed the collapse of the Iron Curtain, a realignment of several nations, and the passing away of the former Soviet Union. The collapse also revealed gross petroleum production practices, abominable disregard of even rudimentary environmental concerns (e.g., 10 percent or more of pipeline oil leaking off en route), and an almost immediate implosion of production levels, by as much as 35 percent, incited by the sustained economic crisis of Russia. Petroleum exports became the only credible marketable product. The nascent petroleum companies created a class of ex-communist capitalists to parallel the assorted shady characters that thrived in the free-for-all, tax-avoiding former USSR. Western companies rushed into these countries with abandon, and some have since become disappointed with the prospects.

China became an oil-importing rather than exporting country in the late 1990s, and its 1.2 billion people supposedly represent the biggest, if not the controlling market for oil in the future. However, the recent Asian economic crisis may suggest that the 21st century will not be Pacific, as widely postulated, but still American.

The Energy Wealth and Poverty of Nations

> "If we learn anything from the history of economic development, it is that culture makes all the difference. ... Until very recently, over the thousand and more years ... that people look upon as progress, the key factor—the driving force—has been Western civilization and its

dissemination: the knowledge, the techniques, the political and social ideologies. … This dissemination flows partly from Western dominion, for knowledge and know-how equal power; partly from Western teaching; and partly from emulation."[1]

The wealth and poverty of nations, the rise and fall of empires and their causes have been studied by economists and historians alike. Toynbee and others have studied history and its rationalizations, while economists have sought to link national behavior with wealth ever since Adam Smith's classic, "The Wealth of Nations."[2]

In 1977, in his remarkable "The Evolution of the International Economic Order,"[3] Arthur Lewis asked, "Why did the world come to be divided into industrial countries and agricultural countries?" The answer was obvious: the former are rich and the latter are poor.

Recently, the world's most notable economic historians, David S. Landes ("The Wealth and Poverty of Nations: Why Some are So Rich and Some are So Poor"[1]) and Charles P. Kindleberger ("World Economic Primacy: 1500-1990,"[4]) attempted to relate the evolution of culture with economic prosperity for the last 500 to 1,000 years. Their books are scholarly works, each running several hundred pages and citing several hundred references (500 in Kindleberer's; more than 1,500 in Landes').

Landes' aim in writing his book was to "do world history … to trace and understand the mainstream of economic advance and modernization." He found elemental reasons for the European economic growth in the availability of property rights, to an emerging middle class and the mobility of people, to places that afforded such rights from places that did not.

In all of these analyses, industrialization is the central and single variable linked to wealth. Even economic decline may be precipitated by industrialization gone astray, as Mancur Olson's 1982 book, "The Rise and Decline of Nations" asserts.[5] He suggests that special interests surrounding wealth-generating industries may become malignant for the entire economy. Wars and social upheavals destroy such entrenched orders, prompting the creation of new wealth from new enterprises.

True enough, but perhaps these notions have already become outdated, or at least incomplete. Energy consumption and access may have already replaced industrialization as the yardstick of the wealth and poverty of nations.

Industries have fled the rich countries, and the process has been ongoing for more than 40 years. The muscular allegories of "heavy" industries so widely espoused by communism have disappeared along with the regimes. The professed industrial prowess in the former communist countries has been discredited by images of shut-down obsolete industries and horrific, unchecked, insidious pollution that filled the news media in the 1990s.

If anything, industrialization, especially of the heavy variety, is today almost antithetical to wealth, passing from the developed countries to the developing world. Understanding the technology and the connections among technology, productivity and wealth is considered far more important today than industry and manufacturing.

Olson's model of industrial destruction and regeneration may be partially true, but, ironically, following World War II, the United States and Western Europe went through de-industrialization and a transition to service-oriented economies. The end of the Cold War offers a more striking case for the overnight collapse of communist-style industries.

The case for energy wealth and its significance is not difficult to make. Entire regions, such as the southern portion of the United States, owe much of their prosperity to the air conditioner, complete with its demanding energy use. Would Houston, Dallas, Atlanta and Miami—America's new cities—have reached their pre-eminence without heavy use of energy in the extreme climate conditions they inhabit (Figure 1)? Not likely!

On the other hand, producing energy does not necessarily imply using it or the associated wealth expected from it. It would be difficult for the casual visitor to believe that oil reservoirs beneath run-down San Tome in Eastern Venezuela produce 1 million barrels of oil per day; by contrast, it is not difficult to see the results of oil wealth in Midland, Texas.

Thus, while energy consumption clearly indicates wealth, holding the technology assets is probably far more critical than hosting the industry itself. Yet, there must be a cultural affinity to technology, one that can also make the connection between productivity or work and wealth. Several authors, still operating under the classic theories of economic history, have alluded to or otherwise ventured to explain the connection between technology (or accessibility), culture and national wealth.[6-8]

Nations and major powers have warred in the past and will go to war for energy resources in the future. (See *Part IV—Red.*) Germany's invasion of Russia

Fig. 1—Houston skyline. Courtesy: Robert Allred Photography

in World War II and the war campaign that famously stalled at Stalingrad, were directed toward the Caspian oil fields. And, of course, the recent Gulf War was not fought for the "liberation of Kuwait."

The energy utilization link to economic growth is precisely why we can assume that China and the Pacific, in general, will be the controlling influences of energy futures and, perhaps, the source of future competition-to-antagonism for Middle East petroleum. This is the challenge to petroleum producing countries: how to manage and take advantage of Asia's increased influence in this most important of all resources. Meeting this challenge may not be easy because cultural mores will have to undergo wholesale behavior modification.

Petroleum Production and Technology in the Cultural Context

The technology for petroleum exploration and production, described in *Part II—Black* and *Part V—Primary Colors*, has unique idiosyncrasies that are not found in any other industry. In addition to the cumbersome physics and the economic dimensions, social and cultural implications are intimately connected with the current makeup of the industry.

To understand the role of technology and culture in petroleum production, it is important to note that "stripper wells" that produce 20 barrels of oil per day, are the norm in North America, while far more prolific new wells that produce thousands of barrels per day are common in several developing countries.

The technological demands of these two extremes are the same: minimal. For the mature reservoir, application of technology may be economically prohibitive; for the newly discovered field, it may not make a difference.

The real challenge lies between these two extremes, where technology and technology management can make a difference. Here is where managing the technology becomes the challenge—a formidable economic challenge and one that is clearly affected by cultural propensity for technology and its deployment. We used the term "technology momentum" in *Part V—Primary Colors* to distinguish this idea from the mere acquisition of technology like pieces in a museum.

The period since the energy crisis of 1973 has created a dominant bipolar situation. Because reservoirs in North America have become more mature, while exploration elsewhere in the 1980s has paid off handsomely, technological innovation has been thwarted. The industry may use vast amounts of technology, but it can coast with existing methods. Matching research, technology needs and economics is always a tricky enterprise; it has become a potentially multibillion-dollar puzzle in the petroleum industry.

Here is the paradox: Although the petroleum industry is technology intensive, and the nature of its operations will surely require more highly advanced technologies in the future, the question is one of timing. When will the newly discovered fields hit a maturity level that requires more than existing technologies?

The multinationals, the traditional primary source of technological innovation, have divested themselves of almost all R&D in the exploration and production areas in response to the industry downturn of the 1980s. The petroleum industry today has the lowest research intensity of all major industries. Naturally, downscaling R&D created a chain reaction, and now, technology proficiency has suffered, and technical people have either fled or been reassigned.

Emerging are a few highly consolidated service companies that offer a smorgasbord of services and technologies in areas that were traditionally the exclusive realm of producing companies. Although a sensitive issue in today's industry, the new players and purveyors of technology have become tantamount to the "emperor's new clothes." What will prevent the new service companies from venturing into the domain of, and competing with, the producing companies? Why would Country X award its challenging concession to a major multinational instead of a technologically more gifted service company?

The Cultures of Petroleum Producing Countries

A handful of countries, including Saudi Arabia, Algeria, Libya, Indonesia, Iran, Iraq, Nigeria and Venezuela, control perhaps 80 percent of all oil reserves, depending on the estimates one uses. They are also by far the world's largest exporters, a role that is almost surely to be enhanced in the future. The maturation of their reservoirs is certain to demand knowledge and technology that is far beyond what is now practiced or available.

The role that technology plays in the evolving stages of petroleum production becomes truly extraordinary because of a common streak that runs through all of the petroleum superpowers: There is little in these cultures to connect productivity (i.e., harder and more efficient work) with wealth. In fact, in these countries, it is predominantly peoples' connections that connect them with wealth and success.

Either tribal leaderships—superficially renamed today with borrowed western titles replete with appellations of "excellency"—or post-colonial aristocracies control these nations.[9-10] They are sometimes interspersed with military strongmen, alternative power holders that often come from the same families.

Connections create oligarchies. In some Middle East countries, societies are created with a multitude of castes and classes composed of locals, Arabs, Europeans and a whole rainbow of servile nationalities.

Are these countries, or are they families?

Such is the environment where petroleum production is the only business. Such is the fabric from which the rest of the world must secure the one product—oil—that is central to the wealth and poverty of their respective nations.

It is natural that tribal leaderships and aristocracies run the petroleum enterprises as national oil companies or all-controlling ministries. In some cases, the company, national treasury and even personal wealth of the rulers have only recently been separated. This separation, though official, may not be obeyed everywhere, perhaps only by form, perhaps still nebulous and corrupt by Western standards (but not necessarily by local standards).

In this setting, the nuances of management and the pretensions of the office become the skills to be coveted, both by those who already belong to the oligarchy and by occasional aspirants from the fringes who long to enter the oligarchy. Hard work and know-how are not the usual mechanisms for advancement. A "buffer" upper class of locals is thus created, not to be confused with the rulers, but clearly to be distinguished from all other classes. With no need to work—in fact, with work almost discouraged—perks and privileges, cars and houses are bestowed. In some countries, unwritten but enforced rules may also dictate the dress.

Who, Then, Does the Work?

In some, more populous countries, non-elite locals are marshaled at pitifully low salaries. In other countries, people from other nations such as India, China, Egypt and Nigeria, and the Palestinians in their diaspora are imported for service. Workers from these countries may not have the latest technological knowledge, but because they have awful employment alternatives, they become the technocrats of the petroleum enterprises. Their culture may be different, their knowledge is often incomplete, and their propensity for improvement is sketchy because real promotion is not possible. They are mercenaries in an environment that neither appreciates their presumed skills nor is willing to adopt them. (See Figure 2.)

The ubiquitous service companies provide the occasional real work. However, because the local management and technical force are not knowledgeable, service companies often lack the necessary supervision and technical cooperation, which can result in inconsistent and substandard quality.

Critical jobs, in which success would be obvious or failure would be precipitous, are often done by Western "lone rangers" who are often misfits themselves, but at least narrowly skilled in the specific art. It is astonishing how many drillers in Libya, a country embroiled for years in official governmental acrimony, came from Britain or even Texas.

Within these petroleum enterprises, the development of "hard" skills and the understanding of technology and its introduction are not high priorities. The work force has either no access to technology or, when exposed to technology, has no motivation to absorb it.

Fig. 2—Chinese workers on a rig

Managers in such cultures can become self-absorbed and whimsical, like a Dilbert cartoon. Acquiring the lingo and the posture of "soft" skills becomes an obvious pursuit. This leads to self-assuring transnational relationships with all too eager, highly paid academics and management consultants. Relating with a dean of a prestigious university is mutually beneficial: The dean gets paid very well, albeit for a short time, and the managers get the necessary boost of confidence in their self-worth by the association. "Memoranda of understanding" are produced by the dozens, often not worth more than the paper they are written on, and having an impact that is even smaller.

In a classic "new lover" syndrome, once a relationship with the foreigners is established, they become less attractive. There is little staying power to absorb the potential benefits from such individuals; there is even less stamina to actually follow up with the implied work.

The priorities of the company management do not escape the minds of engineers and technocrats. Last year, total attendance at all the "hard" technology courses offered by the training arm of a major national oil company was less than the attendance at a single "touchy-feely" course that was devoted almost entirely to management-ese, such as team-building and effective presentation.

Even more troubling, there is virtually no concerted effort toward the management of technology applications.

Cursory and "shotgun" approaches result in extraordinary costs. A fracture stimulation treatment that costs $75,000 in South Texas may cost more than $1.5 million in many major producing countries, where such heavy-investment exercises are performed in safe campaigns, one at a time. Inadequate and grossly delayed introduction of new, but proven, technologies such as horizontal and multilateral wells may mean the loss of billions of dollars in net present value.

An equally telling element is the educational posture of these petroleum producing countries, both with respect to local universities and the importance they assign to the education of their nationals in foreign, more established, universities in the United States or Europe.

Local universities are either long-suffering institutions, often with nebulously defined social, political or local agendas, or recently created "symbolic" universities (because how can a country be a country without one?).

The careers of the graduates are rarely the concern of these universities. Often staffed with inadequately trained faculty—and limping along with pitiful resources and non-existent libraries, laboratories and computing facilities—local universities do, at best, a minimally adequate job. In some countries, the same nationalities that compose the technical personnel in the petroleum company also dominate the professorial ranks. Graduates are, of course, proportionately trained.

In any case, these institutions are rarely the destination of the offspring of the ruling class. They offer neither the prestige nor the four- to five-year entertainment value of a U.S. university, an aspect that is even more important if the student is to return to a conservative society that requires a certain public behavior from its rulers.

A foreign degree is rarely intended to provide the holder with job skills, but instead, to further reinforce his identity. The very selection of who will venture abroad is more often than not restricted to the oligarchy.

A Ph.D. may mean the ultimate in personal reinforcement, even if the skills (presumably gained) are never used professionally. In countries with symbolic universities, a Ph.D. holder with ruling-class connections might serve as an assistant professor for a couple of years, department head for a couple of years (leapfrogging over a number of senior professors from "lesser" nationalities and classes), deputy vice chancellor for a year, and then, minister of petroleum—all within the amount of time that he would need to earn tenure at a U.S. university.

Emergence of Local Services

Recognizing that the service industry has been charged with much of the stewardship of petroleum technology and doing the real work, the service sector has become emblematic of the struggle for autonomy in the national context. National oil companies also pay, and have come to realize that they pay, a premium for their dependence.

Thus, several countries are sincere and enthusiastic about the potential for developing local capabilities and have begun a foray into local services. They are considering ways to spawn a new generation of local service companies, picking one or two fertile areas in which to start.

The North American experience has shown that once local service companies become viable in an area of technology, costs can be reduced by 30 to 60 percent. At 30 percent discounts, the international service companies will play along; at 60 percent discounts, they are long gone.

Of course, no producing country can afford any sustained substandard quality of service or incremental costs, irrespective of the provider, simply on the basis of "affirmative action." And there is no evidence that these producing countries have the cultural predisposition to make the shift. At the same time, some national companies are now aware of the issues involved and have developed considerable sophistication on the subject.

Service maturity or the "ripeness" for emergence of a local service economy can be expressed as a function of two normalized variables: cultural predisposition and technology accessibility.[11]

Cultural predisposition is a broad measure of the competitiveness, cultural beliefs and values that dictate a country's capacity to interact in global commerce, as well as a measure of the cultural propensity for optimizing operations. It may also measure certain national and local traits such as their ability to apply theoretical concepts, or craftsmanship. There are several mechanisms to measure cultural predisposition in the form of questionnaires and polling techniques.[8]

Accessibility to technology is relatively self-explanatory. It is a measure of the overall technological maturity of a region in a spectrum of endeavors, including access to petroleum technology, operational expertise, and legal and financial infrastructures. Emphasis is placed on tailoring technology to local needs and the strongest local technology proficiencies. An emerging local service economy might better serve its region by offering more ubiquitous and less logistically complex services, such as tubular supply, acidizing or cementing, rather than propped fracturing or advanced wireline services.

Figure 3 shows how the service maturity scale technique is applied in assessing the maturity of three important petroleum producing regions: Canada (primarily western), Oman and Venezuela. Ten wide-ranging areas of exploration and production activities are represented. Clearly, the maturity levels of the three locations vary significantly, as indicated by three widely separated points along the spectrum.

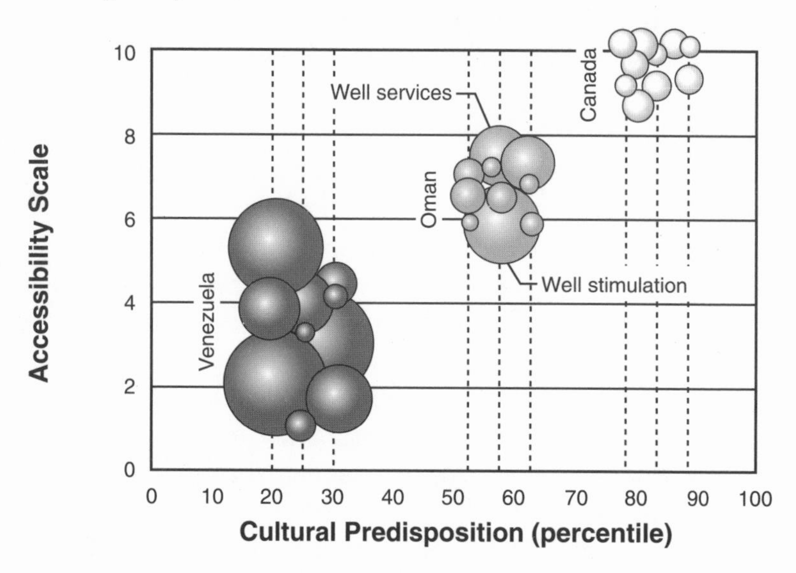

Fig. 3—Application of the service maturity scale

Of special interest are the sizes of the bubbles, each representing the relative cost of an individual petroleum service application. Canada, a mature region in which petroleum services are exceptionally well developed, is represented by bubbles of unit volume. The sizes of the Venezuelan bubbles correspond to the relative magnitude of the cost of individual services, some as much as 10 times the unit volume. Oman shows a middle-of-the-road positioning, significantly better than Venezuela, but considerably below Canada.

If, in the evaluation of a particular service in a specific petroleum region, a large bubble (high cost) is associated with high accessibility or cultural predisposition (preferably both), the region is considered fertile ground for "bursting the bubble."

Clearly, each service does not necessarily enjoy the same accessibility in a particular region. For example, Russians who are well-versed in fundamental sciences and engineering can become readily proficient in reservoir engineering and reservoir simulation, whereas North Americans make excellent drillers and well completion engineers.

Local service companies are not a panacea. For local services to emerge and "work," a country-specific mix will be needed that incorporates national and current business interests along with new players.

The Costs of Culture

Business as usual, defined by culture, may have already caused considerable economic and social hardship. Worse, it may herald disaster for some countries.

Although culture is difficult to change, national behavioral modification can be managed if a company, or even a government, recognizes the problems and takes steps or creates incentives to change it. Among case studies from the past three decades, none is more successful than the amazing transformation of post-Franco Spain, Italy and an emerging muscular Argentina.

The risks for the petroleum producing countries are great, but the potential rewards are even greater. The main message to governments and company management is this: Develop long-term strategies, improve the skills and understanding of the people, create technical ladders, provide true rewards for productivity and true promotion opportunities, and avoid nepotism and paternalism. In other words, provide all the makings of a private enterprise. In all cases, staying power, long-term stamina and discipline are essential.

Technical management should be exactly that. Knowledge and technology should be considered assets to be acquired. As with oil development itself, investments are needed to allow full exploitation of the technology. Opportunistic research and development with local flavor must be given high priority.

More than any patchwork of existing deficient practices, higher education, enormously expensive as it may be, must be the lynchpin of any national reformation. This is the most formidable investment in the future of a nation. Local universities cannot continue to be viewed as contemptuous, orphan stepchildren. True education will bring about not only dry technical skills, but also a mature thinking process and the necessary cultural shift—for the petroleum business and the entire national agenda.

Part VII
Yellow

The constructive and destructive roles of government

President Jimmy Carter at a press conference in June 1977. *(Courtesy: Jimmy Carter Library)*

It was a warm spring afternoon in 1997, a rather ordinary Wednesday, when we got the first call. A reporter called from a prominent national news service and "just wanted to get a quote."

"About what?" was the obvious question.

Later that day, we would get calls from the Bryan-College Station Eagle, the Houston Chronicle, The Wall Street Journal and, eventually, CNN.

Joel Klein and the antitrust division of the U.S. Department of Justice (DOJ) had just given the green light to a large-scale research pact among several major oil companies—Exxon, Mobil, Amoco, Arco, Shell and Texaco—and Texas A&M University. Since we held the key positions as executive director and chief scientist of the upstart coordinating entity called the Global Petroleum Research Institute (GPRI), we were first up to answer the questions and provide quotations.

It was a striking turn, from our perspective, that we would hear of the DOJ's decision from the media. Why would the DOJ announce it publicly before notifying us? The whole atmosphere was foreign and demonstrated to us the chasm between Big Oil and Big Government in the 1990s.

The paperwork that made its way to us later stated clearly, "In accordance with the department's business review procedure, [the no-challenge letter] will be made public immediately." To be accurate, our downtown-Houston legal representatives, the firm of Baker & Botts, got a heads-up fax 45 minutes before the public announcement. But it was lunchtime and the lead attorney was out, so our first notification came from the press.

How was it that the two of us, both long-time and well-known people in the industry, and six top petroleum executives and their attorneys, themselves representing companies worth a half-trillion dollars, did not see this coming? In spite of their massive scale of global activities, neither Exxon, Shell, Amoco, Mobil nor the rest had any substantive history or ongoing intercourse with the U.S.

government—a situation that would only forcibly change with the spate of mergers that started in late 1998.

Our combined knowledge of the government's day-to-day protocol, public as it might be, was almost nil.

We sat around the swimming pool at Michael's West Houston home that evening and pondered a number of questions. How did Big Oil get to be so antithetical to Big Government, while the computer, aerospace and biomedicine industries, the recent flap with Microsoft notwithstanding, were major government benefactors and collaborators? Was this always the case? When did the independent producers, perceived as ruthlessly self-sufficient, become predisposed to government intervention, and when did their interests part ways with Big Oil? Can there or should there be a change?

Thursday morning, our small GPRI staff celebrated the DOJ decision with a champagne brunch at Ron's house and the requisite thumbing of several newspapers.

Then we went back to work. – R.O. and M.E.

⟡

Government—willingly, by default or unwittingly—can turn oil from some of its more constructive colors (green, black, or red, white and blue) to a tawdry yellow.

If democracy and human rights are the cornerstones of modern society, albeit frequently violated, the implicit role of government is to safeguard them, to defend the sovereignty of the nation and to uphold the rule of law. This, in turn, is supposed to define social and economic interaction.

Governments have failed when they have attempted to impose ideological or philosophical supremacy, internally or externally. They have failed when, in the name of egalitarianism, they have deprived their populaces of perhaps the strongest human motivation—the opportunity to gain an edge, be it personal or economic. No matter how noble they are professed to be, all imposed socio-economic systems are likely to collapse, or at least suffer.

From the extreme experimentation of national socialism and communism in Europe and Asia to the most recent populism in Africa and South America, government attempts to stringently regulate human and economic behavior have almost invariably backfired, in certain cases, with catastrophic results.

There is, of course, an economic role for government. It goes beyond the obvious need to show compassion for the downtrodden and disadvantaged. It is more proactive than simply allowing the economy to function freely—a notion considered today to be a mainstay principle of almost all political systems, from the purely capitalist to the socialist and the "Third Way."

First, government can provide a legal framework and a clean slate—free of corruption—for economic activity. This role of government is frequently violated, and the governments of many oil producing countries are prime examples. Although a small percentage of people may benefit from corrupt dealings, nations are never enriched under blatantly corrupt economic systems and lawless political regimes.

Once the rule of law is respected, the role of government can be valuable and constructive in other areas of the national agenda. These areas affect the national infrastructure and critical long-term needs such as the environment, health and energy. Governments, and specifically the U.S. government, have not adequately filled this role in the past and, in most cases, they continue to abrogate their responsibilities.

Instead of policy, governments provide regulations. Although the governmental bodies who impose regulations are often well-meaning, regulations presume an interpretation of the common good and an implicit assumption that business can't police itself. This has taken on a particularly sinister connotation in the petroleum industry.

Regulations, unless imposed as part of a well-thought-out, long-term national policy, stifle the activities of rugged individualists and capitalists. Worse, even in developed nations, regulations can make local industry comfortable with, and then dependent on, government-mandated market reforms. This, of course, is tremendously destructive because it thwarts competition and entrepreneurialism, two of the most important elements of economic success. In the United States, the history of government intervention in the petroleum industry has been a mixture of acrimony between government and entrepreneurs, and the coddling of decidedly inefficient and non-competitive businesses.

In other parts of the world, government policies for the petroleum industry range from self-defeating to plainly disastrous and eminently corrupt.

History of Intervention: The World War I Era

The dust had barely settled from the 1911 breakup of Standard Oil when the petroleum industry was called to perform its patriotic duty during World War I. The United States quickly became a "war collectivism, a totally planned economy run largely by big-business interests through the instrumentality of the central government."[1] Ironically, the antitrust law invoked so dramatically in the case of Standard Oil now forbade such cooperation in private markets.

Private industry was initially fearful of government control, but when the government effectively suspended antitrust laws and liberalized certain provisions of the federal tax code, the industry enthusiastically cooperated with efforts to create what amounted to a U.S. petroleum cartel. A government-industry partnership was sealed, and the industry entered the progressive era of the 1910s and 1920s. Moguls such as Walter Teague of Jersey Standard led the way with "progressive thinking" about industry-government cooperation.

After the war, the industry's new American Petroleum Institute superseded the government's National Petroleum War Service Committee.

A post-war incident cast a shadow over the newfound fraternalism between the petroleum industry and the government. During the war, firms sold fuel oil to the Navy at below-market prices. After the armistice was signed, the Navy demanded that the discounts be continued and even sent warships to pressure coastal refineries into submission.[2]

Nonetheless, the post-World War I period was a good time to be in the oil business in America. An expanding economy (fueled largely by the young Federal Reserve's deliberate currency inflation), automobiles for the masses, and a rapidly improving highway system all led to a boom time in the oil patch.

As always, a bust followed the boom. The bust of the late 1920s and 1930s (distress was evident well before the stock market crash of 1929) was unprecedented in severity. Obvious problems caused by the collapse of the U.S. economy were compounded by new production from several very prolific reservoirs.

Production from the Seminole field, discovered in 1925, followed by the Oklahoma City field in 1929, and especially the gushing East Texas field in 1930, caused the per-barrel price of oil to drop quickly from $3 to $1 to 10 cents. This led to extraordinary actions by the respective state governments—actions that would be incomprehensible today—and extraordinary collusion by much of the petroleum industry.

The premise that selling oil for 10 cents per barrel represented economic waste, and arguments that wide-open production resulted in lower ultimate recoveries, led state governments to prorate oil prices and production based on demand. So-called market-demand proration became the commonly recognized solution for stabilizing the crude oil market. Proration, which had been sporadically ordered during previous periods of surplus, set production limits (allowables) on each well, which varied with oil price and well depth. Tougher laws and enforcement policies were proposed and passed, then embraced by the major oil producers.

But many independent producers had a different view. Not surprisingly, much of their entrepreneurial ingenuity was diverted to circumventing (breaking) the law. Wells were concealed in buildings; crude transports were disguised as moving vans; quarter-acre leases were created; secret pipelines were laid; and enforcement officials were bribed.

Frustrated by the ineffectiveness of proration enforcement, Oklahoma Governor (and presidential aspirant) William "Alfalfa Bill" Murray issued an executive order to shut down the Oklahoma City and Seminole fields in 1931. He imposed martial law and deployed the National Guard to enforce his order until, in his words, "we get dollar oil."[3] Within a few weeks, Texas Governor Ross Sterling, a founder and former chairman of Humble Oil, followed suit in shutting down the East Texas field.

Stability

These actions brought a temporary, if artificial, stability to the market. But this stability was soon undermined by plans to expand the East Texas field. An urgent message sent in 1933 by local producers and the Longview, Texas, Chamber of Commerce to President Franklin D. Roosevelt's Secretary of the Interior Harold Ickes read:

> "Conditions in the largest oil field in the United States … [are] rapidly approaching chaos. Injunctions and counterinjunctions of laws and rules, dissention in the legislature, lack of respect of orders of state commission and widespread violations … are menacing our interests and whole industry and public good by dissipation of irreplaceable natural resources and threatened disorder. Armed men pro-

> tecting leases and pipelines. Pipelines have been dynamited. The organization ... believes that inability of state authorities to bring about effective control of production ... justifies us in requesting immediate federal control of East Texas oil field."[4]

It was not just the independent producers, but even the majors who begged for intervention. Ickes made this comment in 1934:

> "The proposition the oil industry made to the government was the startling one that the government, in effect, take over the industry and run it. It was frankly confessed that the situation was beyond control and that only the strong hand of the government could save it. ... The mental state of these great industrialists can be judged from the fact of their willingness to entrust the destinies of a great business enterprise to a government official who was without scientific knowledge with respect to oil as a product or special acquaintances with oil as a business."[5]

The federal government intervened. The interstate transport of hot oil (oil produced in excess of a well's allowable) became a federal crime, and agents from the Department of the Interior (DOI) were dispatched to east Texas. Within two days, the number of oil tank cars shipped per day from east Texas dropped from 500 to 10.[6]

Private entities such as the Texas Petroleum Council and the Texas Bankers Association became quasi-governmental agencies for restricting the purchase and transport of hot oil.

However, a lower-court injunction put a stop to the DOI actions, and the "hot oilers"—by the sheer force of their numbers and the lack of regulation by agencies other than the Texas Railroad Commission—overwhelmed petroleum production once again. The federal government redoubled by placing an excise tax on wells that produced more than five barrels of oil per day. This single action gave Internal Revenue Service agents the power to inspect wells for illegal production and provided a means of circumventing the injunction against the Interior Department.

While the major petroleum companies were busy cooperating with local, state and federal governments to produce workable market-demand proration, independent producers blamed low-cost oil imports for the industry's price woes, rather than unrestrained domestic production. This was rather creative, since, at that time, the United States was a net exporter of oil, and would remain so until 1948.

Import Duties

Oil had been excluded from the import duties that were imposed on virtually all other goods brought into the United States. The major producers, who had invested in foreign production and downstream assets, were long-time opponents of a tariff on crude oil imports. However, they came to recognize that the tariff was politically necessary for appeasing the independents, and economically necessary for insulating the domestic market so that market-demand proration could work.

Agitation by the independent producers before a revenue-hungry Congress was enough to eventually push through a tariff of 21 cents per barrel, roughly 25 percent of the oil-import price. The tariff caused significant disruption in Venezuela. However, peace had been purchased between petroleum majors and independents, and the latter were brought into the fold of market-demand proration.

By 1936, additional legislation and private enforcement had reduced the hot oilers to minor players, and the crude oil market was finally stabilized—albeit at the expense of consumer interests and foreign relations.

Market-demand proration was soon followed by well-spacing regulations, compulsory pooling and unitization, maximum gas-oil ratios and maximum-efficient-rate production ceilings—all designed to plug "loopholes" in state-regulated market-demand proration, reduce output and improve petroleum prices. Partial state regulation had turned into statewide control, and eventually, it would turn into federal interstate control.

The natural gas industry had been split into three parts by 19th century legislation: production, long-distance transmission and local distribution. The passage of the Natural Gas Act of 1938 was a triumph for distribution and transmission lobbyists. The legislation converted local natural gas distributors and long-distance transmission companies into public utilities, and granted the companies eminent domain, which lowered their right-of-way costs. More importantly, it created captive, protected markets. Producers were neutral or slightly supportive. Consumer interests were once again slighted.

Producers would come to regret their neutrality toward the Natural Gas Act. In 1954, the Supreme Court extended the act to include the production of gas sold into interstate markets. This action was intended to fill a gap in the regulation of the transmission and distribution segments.

The wellhead price of gas sold in the interstate market would be set by government fiat for the next 32 years, resulting in the natural gas shortages of the 1970s.

It is difficult to find a clearer example of intervention breeding more intervention. Producers mounted a concerted effort to neutralize wellhead regulation, but congressional action met with sustained vetoes from both Presidents Harry Truman and Dwight Eisenhower.

MOIP and OPEC

Industry independents were able to push through one more favorable intervention. Eisenhower reluctantly backed the protectionist Mandatory Oil Import Program of 1959 (MOIP) in spite of objections from the majors. MOIP limited imports of crude to 9 percent of domestic demand and created import limits for crude products. National security was the apparent justification for the MOIP, so overland imports from Canada and Mexico were given preference.[2]

Although the MOIP was demanded by the domestic production industry, others smelled opportunity. The "Brownsville loop" was a particularly notorious and creative circumvention. Venezuelan oil was brought by tanker ships to the border town of Brownsville, Texas. There, it was loaded on surface trucks and driven into Mexico and back, then reloaded into oceangoing vessels and shipped to the northeastern United States. The round trip from Brownsville to Mexico served to exempt the oil from MOIP as an overland import.[7]

A direct and ominous effect of MOIP was the formation of the Organization of Petroleum Exporting Countries (OPEC) in 1960. OPEC was created expressly to counter protectionist legislation in the United States.

Although MOIP did not limit imports as much in practice as in theory, it had a sanguine effect on the U.S. industry; domestic production was 29 percent higher in 1968 than in 1959, even in the face of low-cost foreign oil. MOIP remained in place and changed little for 14 years. The implementation of MOIP represented a turning point in the political balance of power in the industry (Table1). Major producers now understood that they could not counter the political clout of the independents.

A far more significant event was not recognized until later: from World War I through MOIP, nearly all legislative government interventions were cooperative—in fact, many were suggested or virtually demanded by the oil and gas industry itself.

A far more antagonistic era was on the horizon.

1970s: The Nixon-Carter Era

The industry felt the first signs of government hostility in 1969 with the reduction of the depletion allowance—that percentage of mineral production exempted from federal income tax. Unfavorable legislation had occurred before, but usually with extenuating circumstances, such as war.

For the first time, the mood in Congress was decidedly against the industry. The primary cause was not consumerism, although that was on the rise as well, but rather regionalism. Oil- and gas-consuming states and their legislators began to feel victimized by market-demand proration, state severance taxes and import restrictions. The always present, but deepening, rift between the major and independent producers was also a factor. It was impossible for government and industry to cooperate in the absence of industry consensus.

In 1970, a Democratic Congress enacted, and President Richard Nixon signed, the Economic Stabilization Act, which granted the president the authority to impose comprehensive wage, price and rent controls. Currency inflation induced by the Vietnam War had, in turn, produced price inflation of 5 percent in 1970. New legislation was the political response, and, violating his previous assurances, Nixon quickly and unexpectedly activated the first federal peacetime price controls. The petroleum industry initially supported the measure as a means of combating inflation.

The program was relaxed over the subsequent two years so that small businesses, such as independent oil and gas producers, were exempted, but the price paid to major producers for their crude oil was frozen at the 1971 price. The majors, who operated most of the refineries, retaliated by refusing to raise the price they paid independent producers for crude. The crude oil price remained essentially frozen while well service and supply costs began to skyrocket.

The states rendered market-demand proration dormant, but domestic supply—with oil prices fixed and oil imports limited—could no longer keep up with demand. Nixon abolished import quotas in 1973 to satisfy domestic demand.

Table 1—History of Government Intervention

Date	Event
1911	Breakup of Standard Oil
1920s	*Progressive era* of industry-government cooperation
1930s	Market-demand proration; oil-production limits (allowables)
1933	Independent producers and Big Oil seek government intervention in letter to U.S. Secretary of Interior
1940s	Intervention breeds intervention: well spacing regulations, compulsory unitization, maximum GORs, gas distributors converted to public utilities
1954	Natural Gas Act of 1938 extended, and interstate gas price set by government fiat for next 32 years
1959	**Eisenhower signs protectionist Mandatory Oil Import Program (MOIP) over objections of Big Oil—separation of interests between major and independent producers and start of antagonist era of industry-government relations**
1960	OPEC created to counter MOIP legislation
1970	President Nixon signs Economic Stabilization Act, leads to freezing of crude-oil prices
1972	"New oil" and stripper wells exempted from price controls; number of stripper wells multiply almost overnight
1973	**First Energy Crisis:** Arab Oil Embargo
1974	Price deregulation to alleviate shortages, except for crude-oil production
1975	Effective "reverse import tariff" hastens slide in U.S. oil production and opens door for OPEC
1977	President Carter forms Department of Energy
1978	President Carter announces massive synthetic-fuels program
1979	**Second Energy Crisis:** Iranian crude removed from market
1980	Windfall Profits Tax enacted
1981	President Reagan terminates oil price controls
1986	**Third Energy Crisis:** Oil price drops from $32 to $10 per barrel in response to surging U.S. production and new oil supplies worldwide

In the final phase of the price controls, "new oil" and "stripper wells" were exempted. New oil was defined as any oil volume produced from a particular property in a given year that exceeded the volume produced from the same property in 1972. Stripper wells were defined as those producing less than 10 barrels of oil per day.

The number of stripper wells blossomed overnight. A well that produced 10 barrels of oil per day suddenly generated more revenue than one producing 20 barrels per day. New wells were drilled to exploit the new-oil exemption. In fact, the drilling boom resulting from these circumventions was restrained only by price-control-induced shortages in drilling rigs and supplies.[8]

Injecting or burning old oil to produce smaller volumes of new oil became an industry joke and an occasional practice.

Shortages of commodities and manufactured goods were endemic throughout the U.S. economy. In 1974, price deregulation was implemented industry by industry to alleviate the shortages. The oil industry was the exception. Import crude oil prices that exploded after the 1973 Arab Oil Embargo were the justification for leaving oil-price controls in effect, even while oil industry suppliers, along with the rest of the U.S. economy, were price-deregulated. General price levels were increasing at an annual rate of 11 percent.

The resulting cost-price squeeze sent domestic oil production into a free fall.

Decades of political favoritism were coming home to roost. In the midst of gasoline shortages and ballooning (but illusory) inventory profits, the public and most of Congress ranked the industry somewhere between common crooks and white-collar criminals. The Oil and Gas Journal told a story of interviewing prospective employees for a clerical position at its Washington, D.C., office. It was not uncommon for job seekers to immediately turn and walk away when they learned that the position was petroleum-related.

Crude oil price controls could not have come at a worse time for the industry or for national welfare. Price controls had the effect of suppressing supply just as national demand was being stimulated. The combination was a godsend for OPEC.

In 1973, Nobel Prize-winning economist Milton Friedman urged the American public:

> "Time and again, I have castigated the oil companies for … seeking and getting governmental privilege. … We shall only hurt ourselves if

we let resentment at past misdeeds interfere with our adopting the most effective way to meet the present problem."[9]

Yet, the U.S. government responded to the oil embargo with a series of faltering measures. The Emergency Petroleum Allocation Act of 1973 was amended by the Energy Policy and Conservation Act of 1975, which in turn was amended by the Energy Conservation and Production Act of 1976. All of these acts were adjusted multiple times during their short lives to address market distortions. This was an obvious exercise in futility since the very intent of the legislation was to create market distortions—those favoring high-cost production (stripper wells, new heavy oil and small businesses) and penalizing low-cost sources.

In 1974 and 1975, U.S. domestic crude oil sold for an average price of $7.27 per barrel, while those exporting oil to the United States received $12.51 per barrel. This amounted to a reverse import tariff of 72 percent (Figure 1). Domestic production continued to slide, and the door for OPEC imports opened wider.

The United States had embargoed its own petroleum resources.

President Jimmy Carter brought an engineer's view to the White House with the premise that the function of the economy could be quantitatively mea-

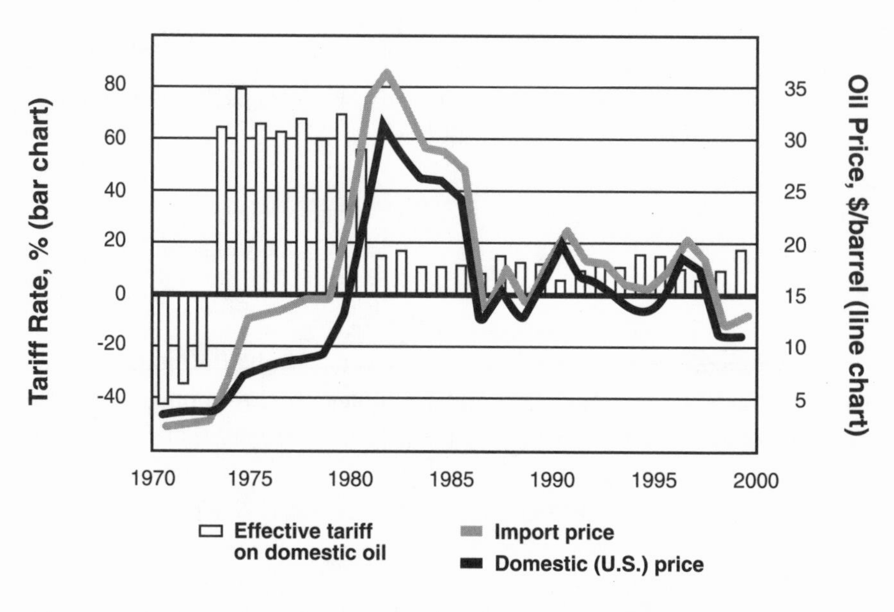

Fig. 1—U.S. domestic and import crude oil prices and effective tariff rate[10]

sured. And because it could be quantitatively measured, it could also be quantitatively managed. In 1977, Carter proposed the establishment of the Department of Energy (DOE) and the adoption of a comprehensive national energy plan based on scientific econometric models. Of course, the scientists who designed these models came with the highest academic credentials and approval.

More realistic, qualitative reasoning was ignored in favor of highly complex computer models of the economy. The main thrusts of the Carter national energy plan were price controls, synthetic fuels and mandatory conservation.

The petroleum industry had trouble extricating itself from a public relations nightmare. Robert Sutton had become the nation's first regulation-generated billionaire by mis-certifying and reselling statutorily underpriced crude. Sutton bragged that he employed a dozen law firms. Eventually he spent 30 months in jail. Ironically, Sutton's trading companies ceased operation in 1981 on the very day crude prices were finally deregulated.[11-12]

The DOE brought overcharging lawsuits against nearly every major oil company, some alleging overcharges of nearly $1 billion. Although the announcement of these suits was frequently front-page news, the eventual dismissal or settlement of the suits rated barely a mention.

The National Petroleum News observed in 1976:

> "A major reason for the public's misunderstanding of the energy crisis and demand for government intervention has been continuing warfare within the ranks of the industry itself. Some oil companies have, from time to time, seized upon opportunities for profit from various public policies. In building a case for government intervention, these companies have also helped to undermine public respect for the oil industry and to bring about federal regulations and controls."[13]

The domestic petroleum industry languished until the 1979 Iranian Revolution removed Iranian crude from the world market. This second energy crisis prompted yet another political response from the White House. Oil prices were decontrolled, but the resulting windfall was to be captured for the American people by the Windfall Profits Tax (first proposed by Nixon).

The 1970s-era government intervention climaxed when Carter delivered his well-known "malaiŝe" speech in July 1978, announcing a monumentally ambitious government-supported synthetic fuels program. Interestingly, the

Synthetic Fuels Corporation was first proposed by Nelson Rockefeller, grandson of the Standard Oil patriarch.

Price signals were muted to the producer by the Windfall Profits Tax, but they were heard loud and clear by consumers. Deregulated crude oil pricing accomplished what a decade of jawboning and mandatory conservation measures failed to do; the U.S. economy became more energy-efficient. Energy consumption per dollar of gross domestic product (GDP) was surprisingly constant from 1959 to 1978. From 1978 to 1985, this indicator improved by 20 percent for total energy consumption, and 30 percent for oil consumption, a remarkable event (Figure 2).

The contemporary downward trend in energy consumption in the United States on a per-GDP-dollar or other energy efficiency basis should not be confused with energy conservation. The fact that U.S. energy consumption is at an all-time high at the start of 2000, at the same time that U.S. energy efficiency is at an all-time high, is indicative of the enduring robust economic expansion that the United States enjoys.

Things were relatively quiet on the natural gas front in the late-1970s. New price-control legislation had little impact because interstate gas price control had been in effect since the early 1950s. However, gas price controls certainly exacerbated the oil crisis as gas became less available and was prone to spot

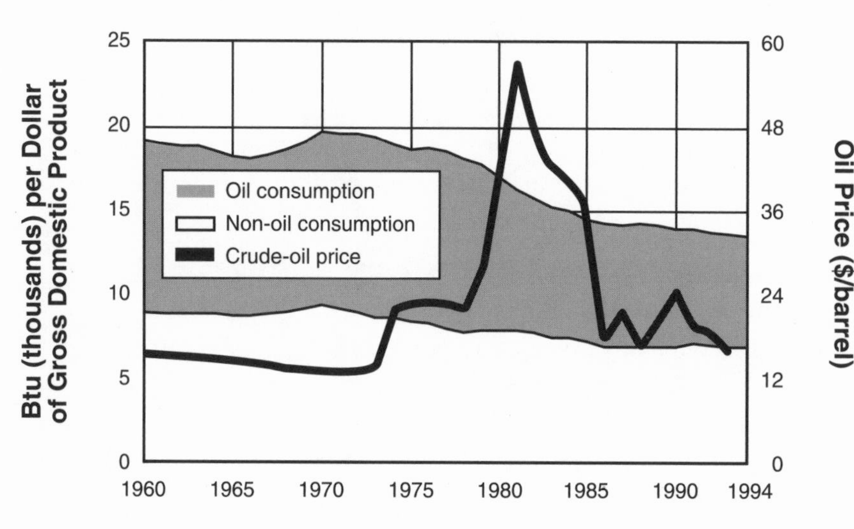

Fig. 2—U.S. energy consumption per dollar of gross domestic product[14]

outages. Industrial and utility gas users with fuel-switching capability tended to opt for fuel oil at precisely the worst time for domestic producers (and the best time for OPEC).

The Carter administration reneged on a campaign promise to deregulate natural gas by enacting the Power Plant and Industrial Fuels Use Act and the Natural Gas Policy Act (NGPA) in 1978.

The Power Plant and Industrial Fuels Use Act prohibited the construction of new gas- or oil-fired boiler installations, in response to the notion that the United States was running out of natural gas—undoubtedly true under the 1954 imposition of wellhead price controls. The NGPA instituted incremental pricing that required industrial gas users to subsidize residential users; introduced a multitude of gas categories, each with its own price ceiling; and extended regulation into the intrastate market.

In 1978, NGPA price ceilings ranged from $1.83 per barrel of oil equivalent to $12.33 per barrel of oil equivalent, all for the same commodity. Needless to say, the exploitation of gas resources took a back seat to the exploitation of the act's price provisions.

This caused one observer to note:

> "A fascinating theme that runs through the long, convoluted history of natural gas regulation is the seemingly inexorable expansion of government intervention. Regulation seems to have spawned further regulation; soon after one regulatory gap was filled, another appeared."[15]

Producers exploited NGPA pricing so effectively that, by 1982, they had begun to price themselves out of the existing market. The deep gas bust of 1982-83 reverberated throughout the U.S. banking industry. High prices of certain gas categories, coupled with take-or-pay purchase contract provisions; shrinking demand caused by the incremental pricing; and a lack of new industrial consumers (under the Power Plant and Industrial Fuel Use Act) created a gas bubble, or surplus, that lasted until the early 1990s.

1980s: The Reagan Era

President Ronald Reagan terminated crude oil price controls eight days after taking office in 1981.

By 1984, oil imports declined 50 percent from their peak under price controls. Falling oil import requirements in the United States and new oil supplies worldwide created a third energy crisis—this time, one that consumers would find entirely satisfactory. Between November 1985 and February 1986, world crude prices fell from $32 per barrel to $10 per barrel as OPEC members, most notably Saudi Arabia, scrambled for market share.

The cartel's pricing power disintegrated almost overnight.

OPEC had underestimated the power of the U.S. domestic energy industry, now unfettered by price controls (as had the econometric models favored by the Carter administration). The Windfall Profits Tax, which was no longer needed in the presence of dramatically declining world crude prices, was soon terminated. The Synthetic Fuels Corporation and the Strategic Petroleum Reserve (SPR) were bolstered by the Reagan administration, but the remainder of the Ford-Carter energy policy was scrapped. Of course, synthetic fuels proved unattractive in the new, low-cost energy era.

One commercial-scale synthetic fuels plant was actually built. The Great Plains Coal Gasification Project produced up to 137 million standard cubic feet per day of pipeline-quality gas from coal. It was built at a cost of $2.1 billion in 1984, with federal loan guarantees providing 71 percent of the total amount.[16-17] The plant reverted to the government when its private backers defaulted on the guaranteed loans, and it was sold to a nearby utility company in 1988 for $78 million (the exact value of the coal rights associated with the plant), mostly courtesy of the U.S. taxpayers.[18]

The Synthetic Fuels Corporation never consumed the remainder of its $14 billion authorized budget.

1990s: The Non-Interventionist Stance

The Reagan era put a neat bookend on the oil industry's government regulatory interventionist story. Since then, the sustained success and real and perceived largess of the industry has justified an almost hands-off approach.

Reagan's successors, often in spite of their public pronouncements, have realized that cheap and abundant energy is the cornerstone of a vital economy. This provides clear rationalization for a diverse set of events, ranging from the U.S. involvement in the Gulf War under President Bush to the unchallenged BP Amoco and ExxonMobil mega-mergers under President Clinton.

Setting aside the poorly conceived and failed Btu tax proposal from early in his first administration, Clinton's most enduring overture toward an energy policy was his 1993 suggestion that the industry rename itself from "oil and gas" to "gas and oil."

The Clinton administration considered intervening on behalf of small producers and has studied the national security implications of rising imports. It is clear that no action is intended or will be taken. Other activities have been limited to marginal energy-efficiency mandates and pronouncements about the future of biomass as a replacement for petroleum—pronouncements that are both unrealistic and destructive to real progress.

In mid-1999, the U.S. Senate approved a bill to provide $500 million in loan guarantees to independent oil producers who were hard hit by the 1998 downturn in oil prices. Legislators in both houses who were from petroleum-producing states opposed the bill. At the same time, but presumably unrelated, an organization of independent Oklahoma producers launched a movement to inflict the legal force of the U.S. government upon producing countries such as Saudi Arabia, Venezuela and Mexico that were accused of "dumping" oil in the United States at prices below their costs. The action was discouraged by the government and broadly opposed by major oil producers.

Both interventionist measures ultimately and predictably failed to materialize, but not before clarifying each player's stance and the underlying historical predisposition. The government will not intervene on behalf of producers; major producers almost uniformly (and sometimes militantly) support this government stance. The independent producers cry foul; and the consumer is oblivious—more likely to complain about a 10-cent increase in the price of gasoline than a trillion-dollar decision, good or bad, made in Washington, D.C.

Such is the government's posture toward petroleum issues at the end of the decade.

The decision to draw down the Strategic Petroleum Reserve (SPR) in 1990 was a sole almost-exception to the hands-off rule and demonstrates a successful application of national energy policy. Speculation and panic-buying on the eve of the 1990 Gulf War sent crude prices to $40 per barrel. When the federal government announced four hours after the first allied bombing attack that withdrawals from the SPR would begin immediately, crude oil dropped to $20 per barrel within 24 hours.[19]

Two weeks later, the announced withdrawal volume was cut in half. Only 2.5 percent of the stored volume was actually withdrawn, but willingness to use the SPR had a dramatic and immediate effect. Strategic is an appropriate adjective for this asset.

The NGPA was phased out between 1989 and 1992. Surviving barely more than a decade, it still proved durable in comparison to the 1970s crude oil intervention.

However, there is a final, farcical chapter to NGPA. Most of the production from the largest United States gas field, Hugoton, was locked into the lower price categories of NGPA. State officials and some (but not all) industry officials cooperated marvelously by determining that after 30 years of production, Hugoton now needed infill wells to drain the reservoir efficiently. Not surprisingly, gas from these new wells would qualify for a much higher price under NGPA guidelines.

A state infill order was entered just as producers were discovering that their experimental infill wells did not add significant new reserves to the field, and just before the phase-out of NGPA was announced. Producers, now wishing to avoid drilling infill wells, were in many cases forced to drill them to satisfy lease agreements that commonly require full property development.

The NGPA was replaced—after much fractious disagreement between producers, transmission companies and consumers—by wellhead price deregulation and a federal mandate to convert gas transmission companies into common carriers. The price of natural gas was allowed to seek its free-market level for the first time since 1954.

Industrial users have returned in force as natural gas consumers, as well they should. Gas is a naturally preferred fuel because of its superior environmental characteristics and large domestic supply.

Government in the New Millennium

The outlook for oil and energy at the turn of the millennium is promising. Regulation of U.S. oil and gas production is relatively light. World and domestic crude oil prices remain near or below the inflation-adjusted historical average price (Figure 3). Domestic natural gas production and consumption are poised to increase at a healthy pace, displacing some oil imports. Price volatility (even taking into account the price crash of 1998) has been declining for several years for both oil and gas.

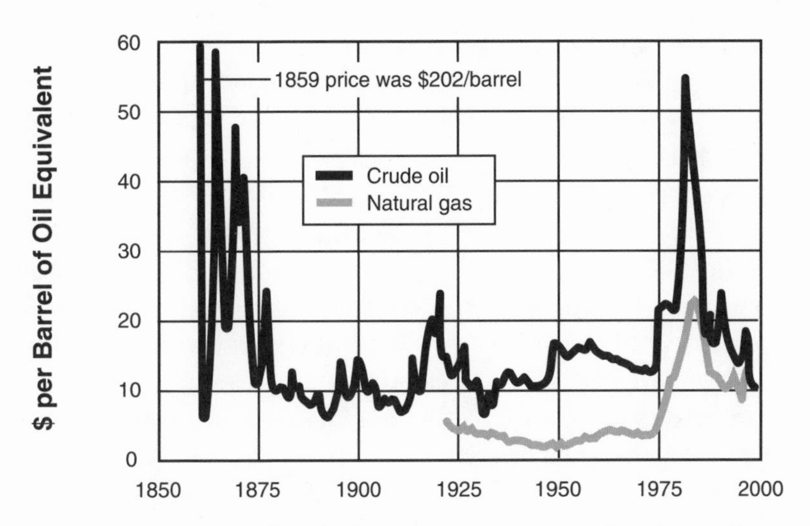

Fig. 3—Inflation-adjusted domestic (U.S.) oil and gas wellhead prices[20]

Future U.S. Energy Policy

One thing is clear, as Richard Vietor summarily concludes:

> "The government's domestic policies for fossil fuels generally failed. They reduced economic efficiency in return for marginal gains in equity, temporary and misleading stabilization of markets, and little or no benefit to national security. For evidence of this, we only need to look to a few key areas of energy policy: synthetic fuels, oil imports, natural gas, and oil price controls."[21]

Government should be a facilitator, creating infrastructure and preserving a level playing field while avoiding legislation that encourages irrational or unlawful behavior by the private sector. This will ensure that long-term needs for the vital hydrocarbon resource are met.

The DOE-forecast increase in the contribution of oil and gas in the U.S. energy mix, coupled with coal's 30 percent, points toward a sustained hydrocarbon dependence of greater than 90 percent through at least the first half of the next century.

Government investments in infrastructure do not interfere with or affect the interests of private industry, and this is good. Tantamount to the establishment of the interstate highway system, the government can have a major and beneficial impact on the petroleum industry. Critical research and development has been de-emphasized by the industry because of the nature of the petroleum business today and the increasingly short-term focus of all industries. Critical long-term research and development in the petroleum industry now demands government support.

A coherent infrastructure policy should include technology investments and information collection and access. It should not include commercialization capital (hundreds of billions of private dollars are available for post-R&D commercial investment) or new taxes (current taxes are burdensome enough).

Although government infrastructure investments have paid handsome dividends in the computer, aerospace and biotechnology industries, government has failed the petroleum industry.

The DOE spent a paltry $116 million of its $18 billion budget (0.6 percent) on oil and natural gas research in 1998, even though these sources constitute 61 percent of the U.S. energy mix and are forecast by the DOE to increase to 66 percent of the energy mix by 2020. The DOE also forecasts a 50 percent increase in energy demand in the same period. The skewed and highly politicized priorities of the DOE have evolved over many years and many administrations, Democrat and Republican, to the point that the potentially strategic DOE is essentially reduced to an expensive garbage dump.

If 10 percent of the DOE budget was redirected to oil and gas research, the resulting $2 billion per year investment would represent a 20-fold increase in petroleum research intensity, and would result in literally hundreds of billions of dollars in economic benefit.

At a minimum, the U.S. government must abandon its currently negative predisposition toward the oil industry—focusing, as it does today, primarily on antitrust and environmental issues—and embrace oil and gas as the ubiquitous driver it is for the economic and environmental well-being of the United States and the world in the next century.

Part VIII
New Green

The politics and environmentalism of oil

View of Earth as seen by
the Apollo 17 crew traveling
toward the moon.
(Courtesy: NASA)

It was quite by chance that I came across a distance-learning video—in Macroeconomics no less—that cited the ongoing battle between oil companies and environmentalists over the development of the Alaska National Wildlife Refuge (ANWR). After all, the modern economist must consider more than just the numbers.

At the mention of ANWR, the tape cuts to a lush forest scene with a small waterfall in the background. A deer grazes in the foreground, and the frame is alive with the sound of birds.

A voice-over narrator explains that economists must also factor into their calculations the importance of saving such pristine places for the children and grandchildren of the world.

The deer lifts its head and lopes off—a very touching scene and one that is hard to find objectionable. But I did.

Only someone who has grown up in Alaska, as I have, would recognize the flagrant deceitfulness of the video.

ANWR is a wide-open, desolate and decidedly barren tundra. There are no forests. There are no deer, only caribou, and in massive herds, not peacefully grazing in isolation. There are certainly no hills for a waterfall to cascade.

Short of oilfield development, there would be no way for a photographer, an environmentalist or anyone's grandchildren to visit ANWR, and outside of oil exploration, there would be no reason for anybody to go there. – R.O.

✧

Latent political or ideological motives, cloaked in pseudo-science, are perhaps the ultimate form of dishonesty. The Worldwide Wildlife Federation has produced television commercial spots with themes that go something like this: A beautiful landscape is shown … then the camera zooms in on a subject, in one case, a female rhinoceros and her baby.

In the foreground, a hunter appears, kneeling and taking aim with his rifle. The picture freezes. Then it takes the shape of a jigsaw puzzle, and the hand of a little girl appears, removes the piece with the hunter, and replaces it with another piece—minus the hunter and his rifle.

The voice-over implores us to give the little girl a better Earth to live on. Fair enough.

Yet, the sequence of a similar TV spot is very troublesome: again, a big landscape is shown. This time, the camera zooms in on a power plant, unidentified, but either nuclear or fossil fuel. The picture again freezes and breaks into jigsaw puzzle pieces. The little girl removes the piece with the power plant and replaces it with one showing three windmills.

Again, the voice-over implores for a better world.

Lost completely is the fact that a typical 2,000-megawatt power plant would need to be replaced by 20,000 windmills of the typical 100-kilowatt capacity, not three. For this imagery to have any semblance of accuracy, the entire landscape (before the camera zooms in on the power plant) would need to be covered with windmills. – M.E.

Environmentalism, couched in difficult-to-combat superficial imagery, has taken a sinister turn. Now highly politicized, it has a gross disregard for the impact that the energy industry has on the world economy. Using moralistic, yet blatantly dishonest slogans and pseudo-science, the environmental movement has digressed dangerously and has replaced some of the most radical movements for social experimentation of the century. One of the most fundamental truths rarely surfaces from the movement: there is no credible alternative to hydrocarbons in either the near or distant foreseeable future.

Modern-day environmentalism—the New Green that stands so tall among the elitist community, multibillionaires and movie stars in the industrialized world—must be distinguished from environmentalism of the stewardship variety, to which we subscribe.

The roots of the environmental movement are both real and ideological.

There is little question that industrialization has brought about visible changes on the planet, especially in comparison to the primitive human state.

Until a few decades ago, these changes—mostly technological, but at times, aesthetic—were the source of pride, an indicator of man's evolution.

To this day, admirers of the accomplishments and monuments of ancient Greece and Rome—and the more recent Italian Renaissance—are also aware that sanitation in these cultures was primitive, that refuse disposal was not a concern, and that life expectancy was less than half what it is today.

The byproducts of industrial progress, such as economic inequality both within and across nations, first came under attack in the last century, and more emphatically in this century. Environmental concerns involving everything from health hazards to sanitation to noise and highly subjective aesthetics surfaced after World War II and became pervasive in the last 25 years.

There is no question that rational reasons forced the new awareness: better understanding of issues from toxicity to radioactivity, spurred by corporate and national irresponsibility of the past. There are notorious cases, such as the Minimata Disease, mercury poisoning in Japan in the 1950s, and the horrifying practices revealed after the collapse of the Iron Curtain.

Yet, the ideology that opposes all industries and industrial development has found a modern manifestation in environmentalism. This ideology, irrational as it may be, often starts from romantic notions of the peaceful nirvana of an agrarian civilization in balance with nature.

Today's movement, often far from being a noble quest, is either a political expression by people of privilege in search of a self-fulfilling cause (but without regard for the consequences), or a campaign by contrarian zealots who hate the energy industry because of its strong capitalist scope. The cause is a convenient rallying post for many wealthy people, members of the entertainment community and pandering politicians. Environmentalism is an issue that simply cannot have an antagonist.

For a billionaire tycoon who probably owns a couple of energy-intensive professional sports teams, it is very inconsistent to exhort others to do without while invoking ill-defined and ill-conceived environmental slogans.[1] When radical environmentalism is applied to developed places such as the United States or the European Union, it almost becomes a caricature, maybe involving the purchase of a $3,000 bicycle for its environmental friendliness. It is another matter and thoughtlessly sinister, however, when it is applied to very poor developing countries.

For others, such as the Green parties, the environment is the rallying cry against capitalism and free enterprise, now that the social engineering systems have failed miserably. (Ironically, few systems have worse environmental records than the socialist-collectivist planned economies.) The devastating consequences to the world economy are not their concern.

Even more fringe groups, self-styled "environmental radicals," may embark into illegal and violent acts and demonstrations in the name of what they think is a higher cause.

For the petroleum industry, whose main purpose is the production of hydrocarbons, the potential for a spill or venting to the atmosphere has always been a cause for concern and has brought fear of real costs and public relations problems. The *Exxon Valdez*, no matter how rare an incident it was, clearly brought forth the reality of the potential dangers and the industry's vulnerability to them.

Anticipation of problems, real or imagined, has affected both the industry and public opinion, although the latter can be greatly polarized among different regions and countries. Offshore development, in Texas, is something to covet; in California, it is anathema. For the industry, environmental considerations have become an important issue in making business decisions about the geographical pursuit of prospects.

Although the petroleum production environment and its potential problems are manageable, and by and large the industry has done a remarkable job in doing so, nothing compares to the discussion of *global warming*. This is no longer a debate on the social responsibility and management of the industry. This is a full frontal attack on its very existence.

The issue is nothing more and nothing less than this: Is the production and use of hydrocarbons, today and in the future, a positive influence on humankind? Or is it, as emphatically stated by U.S. Vice President Al Gore, "the most serious threat that we have ever faced," and our continued use of hydrocarbons, "an effort to avoid facing the awful, uncomfortable truth."[2]

According to Gore, "We should begin with the debate over global warming … it has become a powerful symbol of the larger crisis and a focus for the public debate about whether there is a crisis at all."[2]

Global Warming

A unique and largely unnoticed period of silence reigned in the global warming debate from mid-1998 through 1999. This allows one to view the global warming issue from an almost historical perspective.

Between the two Earth Summits of 1992 and 1997, there was a veritable swarm of publications on the greenhouse effect and global warming.[3-11] The most prolific writers were either environmental zealots or macroeconomists, the two groups that have shaped the debate. Public sentiment was swayed as the scientific "consensus" was touted. Mainstream politicians in many developed nations from the United States to Europe to Australia and New Zealand joined the fray, enthusiastically embracing the rhetoric.

The whirlwind of activity culminated in perhaps the largest convocation of environmental ideologues in Kyoto, Japan, in 1997.

Yet, there was something strange. After an avalanche of publications and announcements leading to Kyoto, practically nothing surfaced in the popular press after mid-1998. Today, presidential candidate Gore commonly gives several-thousand-word speeches and, though the environment is a central theme, he no longer mentions "the most serious threat."

What happened?

First, there is no such a thing as a scientific "consensus." The history of science shows that opinions widely held by the scientific community are often overturned by research and observation. Claims of "consensus" in the global warming debate have been one of its most curious (or ominous) features. The vast majority of global warming pronouncements were based on *models* of catastrophe.

A formidable and growing set of observations now contravenes the consensus.

In mid-1998, at roughly the same time that silence settled over the global warming debate, Dr. Roy Spence, senior scientist for climate studies at NASA, reported the information in Table 1 on NASA's Scientist's Notebook website.[12]

The Intergovernmental Panel on Climate Change (IPCC) estimate for global warming is +0.18 C per decade, or almost 2 C by 2100, measured at the surface. Surface measurements, however, suffer from "Urban Heat Island Effect," basically the effect of robust economic activities in cities. Proponents of surface measurements claim that they have made adjustments for urban warming, but it is difficult to see how this is possible when man-made structures have been added virtually everywhere measurements are taken. Thus, in the

Table 1—Global Warming Predictions

Deep Layer Measurements	
Weather balloon trend (Angell/NOAA)	-0.07 C / decade
Unadjusted satellite trend	-0.04 C / decade
Weather balloon trend (Parker, U.K. Met Office)	-0.02 C / decade
NASA-adjusted satellite trend	-0.01 C / decade
Wentz-adjusted satellite trend	+0.08 C / decade
Surface Measurements	
Sea-surface and land-surface temperatures (U.K. Met Office)	+0.15 C / decade
IPCC Expected Warming Rate	
Estimate of average surface temperatures	+0.18 C / decade
Predicted deep-layer temperatures predicted by climate models	+0.23 C / decade

context of global warming, a discussion of atmospheric temperature changes is necessary. The 1995 IPCC estimate for "deep layer" atmospheric temperature change, based on climate *modeling*, is +0.23 C per decade.

There are two methods of *observing* atmospheric temperature change directly: weather balloons and satellites. Satellites, in use since 1979, provide a uniform global picture of the temperature of the lower atmosphere and provide the most accurate picture currently available. Weather balloon data suggest a global *cooling* trend in the range of -0.02 C to -0.07 C per decade. The raw temperature data collected by satellites suggest a cooling trend of -0.04 C.

Recent efforts have focused on calibrating the satellite readings for orbital decay or "downward drift." Using corrections published by Wentz and Schabel in the Aug. 14, 1998, issue of Nature,[13] a trend of +0.08 C per decade replaces the uncorrected trend of -0.04 C per decade. Corrections performed later by Spencer and Dr. John Christy (The University of Alabama in Huntsville) suggested a slight cooling trend of -0.01 C.

A National Research Council-sponsored review of the satellite temperature-trend controversy will be published in early 2000. Dr. Spencer and his most vocal critic both sit on the review panel. The report will again, mistakenly, try to accommodate all views and establish a consensus. The much more legitimate process of scientific publications and peer-reviewed work lags well behind the political and populist venues.

Suffice to say, the IPCC predictions now appear highly inconsistent with a fairly robust set of observations.

We do not particularly debate that the globe today might be warming. It does appear that the earth may be warming as part of a 100-year trend (much of this trend predating World War II and the growth of CO_2 emissions). What we debate is the highly politicized and pseudo-scientific process that is not even internally consistent.

It has been suggested that there was an abrupt 16-degree warming at the end of the last ice age.[14] Clearly, this event was not caused by an SUV-driving prehistoric man.

What's Next?

Will an obvious setback in the global warming rhetoric tone down the environmental movement? Facts have never really been important in zealotry.

A major United Nations environmental report that was released in September 1999 hints at the answer.[15] The lengthy executive summary, titled "Major Global Trends," does not once mention global warming and makes only two perfunctory references: one to "climate change" (positive and negative temperature changes included), and a second to increased CO_2 emissions. In the absence of global warming, CO_2 emissions are more of a virtue than a vice—CO_2 is the main substance that promotes vegetation growth!

Rather, the U.N. report claims that "new dimensions have been added," and describes other environmental catastrophes that are in the making: a global *nitrogen* problem, forest fires, and increased frequency and severity of natural disasters. The latter certainly has no industrial or human cause.

The same report concludes: "A tenfold reduction in resource consumption in the industrialized countries is a necessary long-term target if adequate resources are to be released for the needs of developing countries."[15]

An update of the United Nations IPCC's 1995 report is due in 2000, and with a $1.7 billion climate change research budget in the United States alone, for certain there will be much to say.

The Problem

Beyond the nuisance factor, environmentalist slogans and activities pose no real long-term danger to the petroleum industry. No multimillionaire environmentalist can truly live through an energy shortage, nor can any politician survive a self-imposed energy crisis.

Petroleum is the lifeblood of current and emerging world economies. Its use will grow, both in developed and developing nations—it is the only way. Rhetoric notwithstanding, it can be done in an environmentally prudent fashion, and we believe that it will.

The problem is one of public perception, and the counterproductive-to-devastating force that it can be when it gets off course. Public perception can take a perfectly logical course of action and turn it inside out.

Chuck Colson describes a cartoon that shows a buxom woman sitting in the witness stand in a courtroom.[16] The judge declares, "There's zero evidence that silicone breast implants cause disease!" Then he orders, "Give the plaintiffs three [billion] dollars anyway!"

According to Colson, who is making a plea for simple justice, a National Academy of Sciences panel has concluded that there are indeed some risks associated with breast implants, like infections caused by leakage, but that there is no evidence that the implants damage the immune system or cause disease. Clearly, the facts emerged too late to help Dow Corning, the largest maker of breast implants. In 1998, Dow paid a $3.2 billion settlement to 176,000 women who claimed that they were injured.

Why did Dow pay? According to Colson, "The company knew that in today's cultural climate, it didn't stand a chance."[16]

It is not likely that anyone is going to correct the campaign of global warming and other misinformation promulgated by the environmental ideologues. In fact, they are likely to attempt to translate their missteps of the 1990s into political power in 2000.

For sure, the debate will not suddenly become transparent and steeped in integrity, for several reasons:

- Governments have embraced the environmental zealot movement and shaped far-reaching legislation around it (Australia is a prime example).
- Politicians have staked their careers on global warming as a fact, and they will be hard-pressed to change their stands in mid-course.
- A whole generation of school textbooks has been written and disseminated and will not be recalled.

As in the case of Dow, intellectual dishonesty by a very clever and financially well-endowed generation of environmental ideologues will continue to prey on the intellectually lazy or morally weak whenever it can (Figure 1).

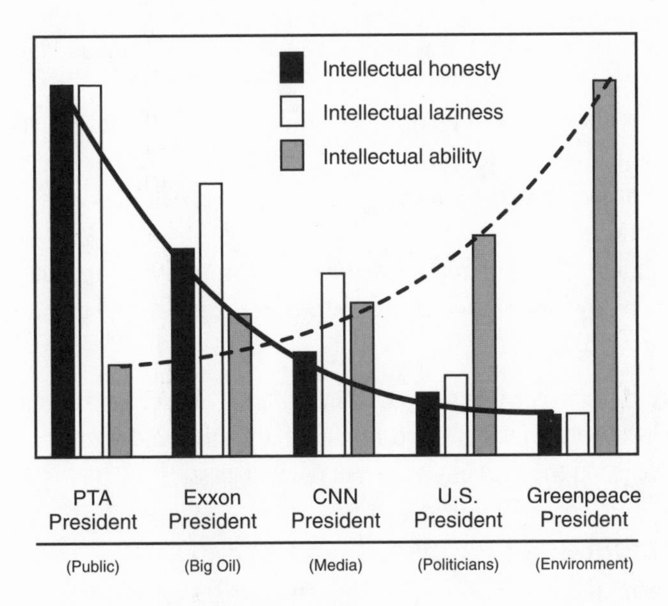

Fig. 1—The Environmentalist Prescription

Figure 1, though perhaps offensive to some, seems to reflect a large consent in the public mind. It is surprising how even leaders in the debate can be blissfully ignorant of the most rudimentary fundamentals behind global warming.

Consider, for example, this 1998 statement from Dick Morris, former political adviser to President Bill Clinton and would-be adviser to Al Gore:

> "I think there is something going on [with] the environment and global warming and all that. There have been just enough incidents of crazy weather around the world to make people think maybe the environmental nuts and tree-huggers are right. I'm going to check it out with a poll and show Gore the numbers."[17]

Pseudo-science and deceit have become mainstays of the environmental debate at the turn of the millennium.

A Better Way

Many industrialists and economists have espoused environmental stewardship rather than environment as a religion, and despite the gross distortions of certain environmental propaganda, their record is quite good:

- Tree growth in the United States today exceeds the commercial harvest by 37 percent.
- Voluntary advances in farming practices since 1940 have saved nearly 1 million square miles of land that would otherwise have been plowed under for food production. In fact, lots of old farmland is being returned to forest.
- The amount of carbon reabsorbed in North America today matches fossil emissions, owing to modern agriculture, fertilizer use and the world's best forest coverage.

According to a recent Forbes article, "America has reached a unique stage in its development. Alone in the world, we're reforesting and farming enough to suck more carbon than we blow, current data suggest."[18]

Environment and Energy

The petroleum industry has come a long way from the exploitative and abusive days of Spindletop. The current generation of petroleum engineers, executives and oil-field workers are quite environmentally conscious. The authors of this book started an environmental consulting and cleanup firm in Southeast Asia in 1990; today the company employs 45 people and has offices in five cities.

While major petroleum companies take different approaches—BP, playing the press endlessly; Exxon, maintaining a strict press-relations-unfriendly demeanor; and the late Mobil, advertising in popular weekly news magazines for a saccharine public persona—they all share a self-imposed vigilance that applies to their environmental behavior as well as their social responsibility. They strictly enforce their so-called "international standard" everywhere they operate, often in spite of the objections of host governments who would rather avoid the unnecessary expenditure.

This behavior, especially among their top executives, would be out of place or even career-ending at, for example, the Walt Disney Co., Coca Cola and, of course, Nike.

For the record:

- Major multinational oil companies today are the foremost purveyors of the so-called "international standard" for safety and environmental performance.
- Today's new cars produce 95 percent less pollutants than new cars in the 1960s, thanks to cleaner-burning gasoline and the latest car-emission technologies.
- Surface disturbances caused by oil-well drilling (the drilling "footprint") have shrunk dramatically with the advent of better seismic mapping techniques, horizontal wells and underground gathering systems.
- There has been no major spill from an offshore production facility for nearly 25 years, even though more than 25 percent of U.S. production is obtained offshore.
- In response to the *Exxon Valdez* accident, the industry is quietly but deliberately shifting toward double-hull tankers, in many cases, well ahead of the federally mandated timetable.

Environmental stewardship in the oil industry, can, has and must continue to keep pace with the uncovering of real environmental information, and this can happen with rapidly advancing technology.

Rather than a head-to-head battle between "Big Oil" and "Big Environment" and the win-at-all-cost mentality that has emerged, we advocate working progressively toward solutions. Emission trading in lieu of zero-tolerance legislation, for example, creates incentive for progress that is constructive instead of destructive.

The transition to natural gas as the primary fuel and the very real future of fuel cells may eventually become a common ground for all well-meaning people on both sides.

The New Energy Economy

Much ado has been made of fuel cells and the emerging hydrogen economy.[19,20] It is certainly insightful and compelling to notice the successive use of fuels with ever-lower carbon content, starting in the 1800s and running through to today—from wood (high carbon content) to coal, oil, natural gas and envisioned (no carbon) hydrogen. Each fuel has been cleaner and more energy-intensive than its predecessor. Each has also been more technologically sophisticated.

Ironically, natural gas will be the fuel of the hydrogen economy (really a fuel-cell economy). A careful check of the fine print in the advertising for a new fuel-cell electricity generator reveals that the fuel cell runs on hydrogen that—the advertisement adds in parentheses—is extracted from natural gas or propane.

Natural gas is the common denominator among a flurry of conflicting opinions on oil prices, energy supply, fuel cells, global warming and economic development. It will foster all reasonable wishes of environmental groups while supporting major and sustained economic activity throughout the next decades.

Here is the scenario for what we believe will happen in the United States. Oil (and gas) companies will produce massive volumes of natural gas from deep offshore Gulf of Mexico. The necessary lease rights are in hand, and natural gas is already on an unstoppable path to become the fuel of choice for space heating and power generation. The not-so-secret multibillion-dollar R&D efforts and alliances among car makers (General Motors, DaimlerChrysler and Ford) and major oil companies (Exxon, Royal Dutch/Shell and Texaco) suggest that fuel-cell-powered cars are on the immediate horizon. This will happen much sooner than most people think, and not because of off-the-wall opposition to the gasoline engine.

Carbon dioxide emissions will drop dramatically as a result, regardless of whether global warming is fact or fiction.

All of this will, in the future, leave liquid petroleum to be used for its far more valuable purpose as a feedstock in the manufacture of synthetic materials like plastic milk jugs and pantyhose, as it will no longer be "wasted" on the generation of primary energy or for transportation fuel.

If this sounds too rose-colored to be true, consider the following.

First, natural gas is cheap and abundant. The energy content of a standard unit of gas (thousand standard cubic feet, Mscf) is about one-sixth of the energy content in a barrel of oil, while its price has historically been one-eighth the price of oil or less. With gas selling at $2.50 per Mscf as it did in late 1999, the break-even price is less than $15 per barrel—a bargain when oil was selling for $27 per barrel.

Second, the amount of carbon emissions from the straight combustion of natural gas is half that of oil and coal for the same energy output. Gasoline and diesel derived from gas-to-liquid conversion, the most prosaic of advanced uses, have far less sulfur, nitrogen and complex aromatics than conventional gasoline. The process is economically competitive with refinery gasoline derived from $20-per-barrel oil.

Third, there is a clear momentum toward natural gas in the most influential of all markets, the United States. The fraction of oil that goes to power production has already dropped to less than 5 percent. With deregulation, nearly all new power plants will be gas-fired (cheaper on a per-megawatt basis), which will cut into the market shares for coal and nuclear power. These power plants will be fed by the massive gas reserves in deep offshore Gulf of Mexico (as the production economics improve) as well as the highly developed and prolific gas provinces of Canada.

Very large and known gas supplies in Alaska, currently reinjected or dormant, can be brought to the market, either as liquified natural gas, as the product of a gas-to-liquid conversion or, far more likely, directly through a pipeline that connects to the continuously expanding Canadian pipeline system. Such a project has been mentioned often in the past and its likelihood is improving with time.

Fourth, the underdeveloped international gas market is coming to life. Algeria, Qatar, Oman, Saudi Arabia and Venezuela are all shifting emphasis to gas. With its oil reserves seriously depleted, Trinidad is putting together a formidable gas industry. Qatar has initiated several gas-to-liquid and liquified-natural-gas deals in a progressive movement to bring its 250 trillion standard cubic feet of natural gas to market.

Of the G-7 countries, Canada, Germany, the United Kingdom, Italy, France and Japan are poised to expand their gas use greatly. Together, with a combined population that is 1.5 times that of the United States, these countries consume only half as much natural gas as the United States (11 trillion standard cubic feet per year compared to 22 trillion standard cubic feet per year).

The Drawbacks

The largest drawbacks to natural gas are political. Building and facilitating natural gas infrastructure and the adroit use of taxes and tax incentives can play a very constructive, substantive role in a social and economic transformation during the 21st century.

The transformation to a natural gas economy, an issue that is explored further in *Part IX—Purple*, will not be painless. Supply and demand will struggle to stay in balance as gas consumption increases; the struggle will be complicated by our inability to predict seasonal weather anomalies and by the masking effect of gas storage. Over a period of several years, and depending on the eventual manifestations of El Niño turned to La Niña, there will be at least regional shortages of natural gas. Natural gas shortages can be very acute, as demonstrated when the natural gas price in Chicago spiked to more than $20 per Mscf in February 1996.

The obvious way to mitigate short-term imbalances in a natural gas economy is with imported liquified natural gas, or LNG. Enron's gas position in Trinidad could become quite strategic, for example, based on its proximity to the United States.

While the global LNG trade has grown significantly during the past 30 years, the United States has been on the sidelines, and justifiably so, based on the prevailing economics. When natural gas shortages surface in the next several years, the economics for LNG in the United States will change almost overnight, but the time required to construct new LNG tanker ships and vaporization facilities (necessary to employ the supercooled LNG as it is offloaded from the tanker) will be excruciating.

In 2000, there are only two active LNG receiving terminals in the United States: one near Boston and one in Lake Charles, Louisiana. The combined vaporization and trucking capacity of these facilities is 1.3 billion standard cubic feet per day, or 2 percent of U.S. daily gas consumption.[21-22] Even with the 1 billion standard cubic feet per day of LNG import capacity that will probably be activated at Cove Point, Maryland, the shortage of facilities will eventually manifest itself as a choke point in the U.S. system.

This is an area where the federal government could play a significant role in building infrastructure that no single industry player will be able to justify until it is too late.

Instead, the Clinton Administration in 1999 was making statements that "biomass will be for the next century what petroleum was for this century."[23] Clinton said he wants to develop fuels from plant and agricultural wastes. Although this sounds like a great idea, a more practical possibility and the one that industry is focusing on is the production of biomass energy from purpose-built tree plantations that are "sustainably managed." Trees would be replanted to absorb the carbon dioxide liberated during combustion.

This scheme meets the sustainability requirement, but it fails to meet a couple of others. First, it represents a remarkable step backwards, to the age of cutting down trees and burning them to meet energy needs. It is difficult to imagine environmental groups suddenly advocating the cutting down of any forest, even a purpose-planted one. Second, the logic of this type of biomass *vis-à-vis* global warming is identical to that of hydrocarbons. The carbon dioxide generated by the combustion of fossil fuels in North America today is already being absorbed (more than 100 percent) by means of longstanding major reforestation and farming activities.[18]

A Final Word

Everyone, from vociferous environmentalists to hard-nosed oil company executives and anybody in between, can rally around the potential benefits of natural gas. Yet, the related discussion of LNG prompts a somber reminder; oil and gas are potent and dangerous products. Natural gas, regardless of its environmental benefits, is not benign.

If a 20-foot hole was ripped in the side of an LNG tanker while it was docked in Boston Harbor; and if there was a moderate southwesterly breeze; and if several hours passed before the escaping product ignited, the blast would level buildings on Long Island. A large explosion in Cleveland in 1953, caused by a relatively small volume of LNG, provides a vivid precedent for this danger.[24]

Part IX
Purple

An industry for the third millennium

New York skyline at night.

The Acropolis of Athens and the monuments that it encompasses are the finest examples of classical Greek architecture. They are symbols of the confluence of thought, culture and art that have characterized the fifth century B.C. as the Golden Era, not just of Athens or Greece, but also of the world.

I remember, as a Greek child growing up in Cyprus, learning about the importance of the Acropolis and what it meant to the history of the country—a history measured in centuries, where 500 years was only yesterday.

Of course, my history lessons were very selective, always downplaying less glorious events whose remnants were evident around the Acropolis. I was shocked the first time I visited the Acropolis and Athens during my early high school years. The area was in an obvious state of disrepair, and there was rubbish everywhere. Modern Athens, perhaps one of the ugliest and most unorganized European capitals, provided a sorry backdrop. Air pollution and its more formidable byproduct, acid rain, were wreaking irreparable damage to the marble monuments. The faces of the Caryatides, the beautiful maidens supporting the roof of the Erechtheion temple, were losing their features. The only statue that was keeping its facial features—to the chagrin of all Greeks—was the one in the British Museum in London, taken by Lord Elgin two centuries ago.

It was not difficult to see that something needed to be done—urgently.

With large grants from UNESCO and others, successive Greek governments embarked on a very ambitious plan for reconstruction and reclamation. The Parthenon marble and the statues were given protective polymer coatings. More recent structures were brought down, and eyesores in the surroundings were removed. A former minister of culture, actress Melina Merkouri, declared that the intention was to make the area look and feel as "it was."

The results, still evolving, are stunning. I have watched the facelift over the years during my regular visits, and the Acropolis is re-emerging, weathered but ageless.

And yet, in this purging of everything new that would mar the ancient glory, the Greek government wanted to give both locals and visitors a spectacular night view. Today, the Acropolis, bathed in light, can be seen from everywhere in Athens, a city of four million people. The vision, ethereal as it may be, is the result of many powerful lights, perhaps a couple of million watts.

Electricity and the light it provides does not seem to figure into the conflict of the old and the new. Without it, the monuments of the Acropolis would still be in the dark. – M.E.

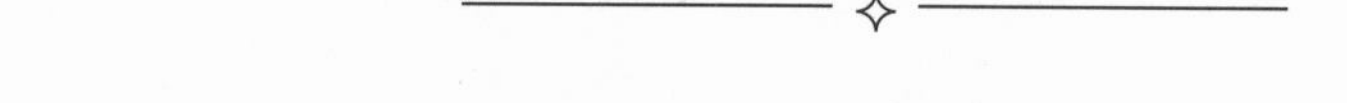

The future of the world that we know is also the future of energy. This future may not be limited to the energy sources we know today. One thing, though, is for certain. There is no room for reductionist ideologies, prompted by false senses of fairness, or for calls for voluntary transnational reallocation of natural resources. Energy use is so vital, so seminal to our well-being and quality of life, that "wealth through energy" should be the mantra of the world.

Energy is the crowned king, and oil is now the color purple.

This is the reality, but the sociopolitical dimensions of energy are far more complex. The world is more optimistic than ever in facing the future, but there are still many voices of discontent and dissonance.

First, there are the romantics who believe that any lifestyle that brings us away from their vision of natural (read primitive) man is somehow a rebellion against nature. They are matched fittingly with the doomsdayers and their earth-destroying global warming. Then, there are those who are guilt-ridden for having and hoarding something that others do not have. There are also the internationalists such as the authors of the U.N. Global Environmental Outlook 2000 report, who are asking the developed world to reduce its use of natural resources by 90 percent (!) to give the developing world a chance to catch up.[1]

These ideologies—spanning natural resources and just about anything that affects nature, from water to pesticides to medicines—are widespread with varying degrees of naiveté, stridency and discord. They have shaped national and international debates and generated social conflicts. Energy, especially oil and gas, brings out all of those emotions and more. Why? Because presumably knowledgeable people believe that reliance on such depletable resources simply intoxicates us and will render us unfit when the resources run out.

We have a tenet that stands in sharp contrast to this one.

Far from any catastrophic and self-defeating notions of reducing energy consumption, and far from any civilization-ending, apocalyptic side-effects, energy use should, in fact, be expanded in the future. Resources should become more accessible, and all possible alternative and advanced energies should be marshaled. The world's top priority should be to secure energy sources indefinitely and to employ technologies to render them at low cost. Cheap and abundant energy is key to the well-being of humankind.

This, bar none, is the most populist and humane vision for the next millennium.

Where Are We Now and Where Are We Going?

Today, at the end of one millennium—whose last century both shaped and was shaped by energy—and at the beginning of a new one, energy consumption is lopsided among nations, as shown in *Part I—Green.*

It is essential to understand the makeup of today's energy sources, both hydrocarbon and non-hydrocarbon, and even more important to understand their future. Figure 1 shows the energy mix of the past and a forecast of the world energy mix for the next 20 years.[2,3] The world consumes 400 *quads* (quadrillion British thermal units) of total energy per year. This is equivalent to 200 million barrels of oil per day. Of this, 40 percent is oil, 22 percent is gas, 24 percent is coal, 6 percent is nuclear, and 8 percent comprises all other energy forms, mostly hydroelectric. (Wind and solar, combined, make up 0.5 percent of the total mix.)

The United States' 280 million people (4.6 percent of the world population of 6 billion) consume 25 percent of all energy. Simple arithmetic can show that U.S. residents consume seven times the per capita use of the rest of the world, including all developed and developing countries.

Yet, traditional and superficial claims of energy waste and resource hoarding by both friend and foe are, and should be, out of vogue. The United States employs energy to generate wealth, and does so in a highly efficient manner. In the competitive and free economic environment, a company that uses energy wastefully probably will not survive.

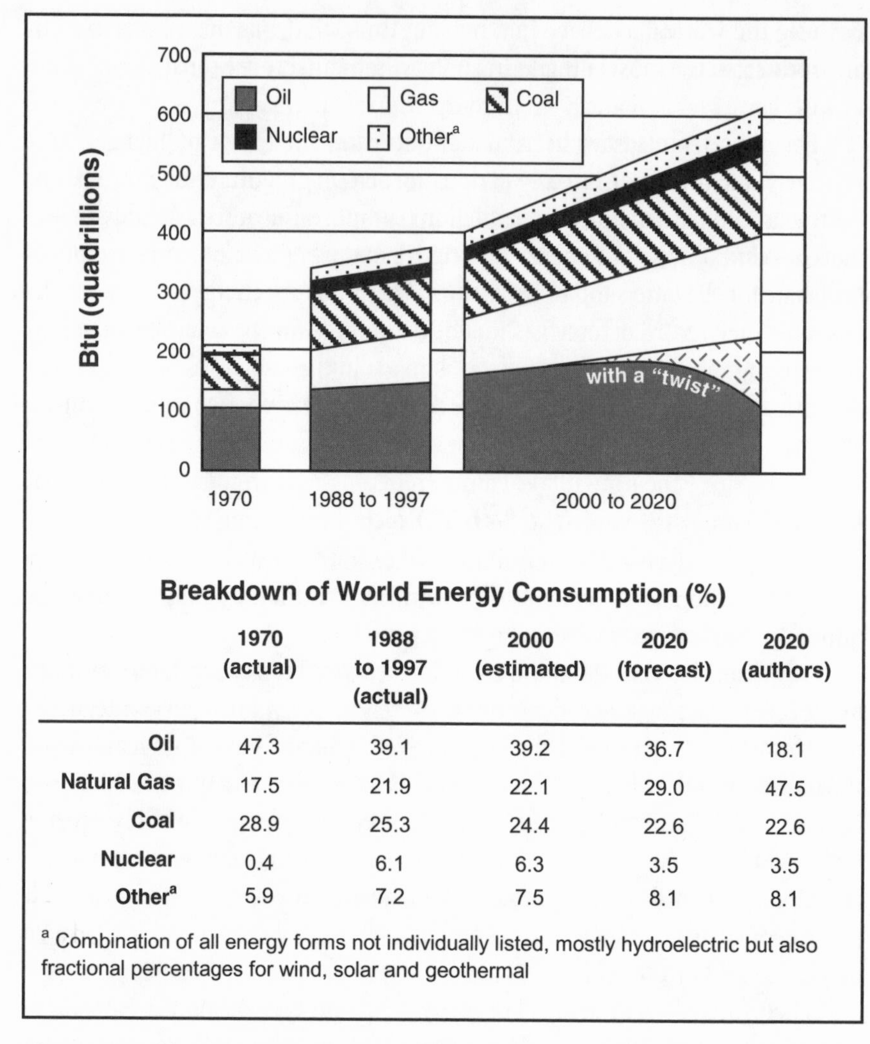

Breakdown of World Energy Consumption (%)

	1970 (actual)	1988 to 1997 (actual)	2000 (estimated)	2020 (forecast)	2020 (authors)
Oil	47.3	39.1	39.2	36.7	18.1
Natural Gas	17.5	21.9	22.1	29.0	47.5
Coal	28.9	25.3	24.4	22.6	22.6
Nuclear	0.4	6.1	6.3	3.5	3.5
Other[a]	5.9	7.2	7.5	8.1	8.1

[a] Combination of all energy forms not individually listed, mostly hydroelectric but also fractional percentages for wind, solar and geothermal

Fig. 1—World energy consumption superposed with authors' forecast of gas market share of oil starting in 2010

In *Part VI—Colors of the Rainbow*, a strong case is made for energy consumption as today's prime indicator of the wealth and poverty of nations. For the rest of the world to start bridging the gap, let alone catch up, there is a need for massive mobilization of readily available resources and the recruitment of others that are underutilized or not used at all. This will be like nothing the world has ever seen, nor anything that has been fantasized by anybody, from the most ruthless Darwinian capitalist to the most ardent egalitarian.

Several attempts have been made to forecast the future of energy over a relatively safe time span, say 20 years, using annual growth rates for predicting both total energy demand—low, medium (graphed in Figure 1) and high scenarios—and the various energy sources. Presumably, a blanket rate is not appropriate for the latter, and other factors and trends are used.

The track record of forecasts made 20 years ago for the evolution of energy consumption is respectable. Predictions made in the late 1970s of energy consumption levels in the 1990s were quite accurate.[4] This is remarkable, considering international economic recessions and booms and, in the case of oil, the price collapse of the mid-1980s. Painful repercussions from the price collapse are still felt today by some very large petroleum producing countries such as Algeria, Indonesia, Nigeria and Venezuela.

The "market share" of energy sources is another matter. Political, economic and technological events affect the energy mix.

Although oil forecasts for the United States were reasonable, forecasts made in the late 1970s that nuclear energy would reach 11 or 12 percent of the energy mix by 1990 were not met. Real and contrived influences led to an actual share that was less than half of the forecast percentage, and it remains near that percentage today. Nuclear energy's market share is not expected to increase in the near future and, if anything, it is likely to decrease. By 2020, it is expected to contribute less than 3.5 percent of the total energy mix.

A similar fate was met by coal. Projections in the late 1970s suggested a 26 percent share by 1990; the actual market share was right at 20 percent.

The primary event was, of course, the difficult-to-predict deregulation of natural gas wellhead prices and an eventual U.S. mandate that forced the conversion of gas transmission companies into common-carriers (providing "unbundled" services). This sequence, followed by the latest initiatives to deregulate electricity, has propelled natural gas to be the premier growth energy industry in the country.

Figure 1 shows both the conventional forecasts for world energy consumption to 2020 and our own forecast with a considerable twist.

First, the conventional forecast:[2,3] Total world energy demand will increase by 2.1 percent per year, reaching 612 quads or 300 million barrels of oil equivalent per day by 2020. Oil demand will increase by 1.8 percent per year, but its share of the overall energy mix will drop slightly from 40 percent in 2000 to 37 percent in 2020. Gas use, on the other hand, will increase by 3.3 percent per year, and its share will escalate from 22 percent to 29 percent during the same period.

We think this forecast disregards the huge potential of gas, discussed extensively in *Part VIII—New Green.*

Now, our forecast: By 2005, the age of hydrogen will dawn—pushed first by fuel cells running on natural gas or natural gas liquids and gradually expanding everywhere in both industry and life. Therefore, a more logical growth rate for the use of natural gas starts at 1.5 times the conventional forecast, accelerates to 2.0 times in 2005, and then climbs to 2.5 times that predicted by the conventional forecast, starting in 2010. By 2020, we predict that natural gas will make up a commanding 45 to 50 percent of the worldwide energy mix. Our forecast suggests that the higher-than-expected growth in natural gas demand will be offset by an equivalent reduction in the demand for oil, reducing oil to less than 20 percent of the overall energy mix by 2020 (Figure 1).

It Will Not Be Easy and It Will Cost

Over the next decades, producing oil at the pace that the world is demanding is not going to be a snap. Nor will it come without cost by just "opening the taps."

At the start of the new millennium, the world petroleum demand is 75 million barrels per day, and over the last decade, it has increased at 2 percent per year. Even with the Asian economic crisis, which contributed to the oil price collapse of 1998, petroleum demand still increased, albeit by less than 1 percent. Based on the U.S. Department of Energy (DOE) mid-range "reference case" forecast,[2,3] which assumes a 1.8 percent annual increase in oil demand for the next 20 years, the daily oil consumption should be 90 million barrels per day in 2010.

At the same time, existing production, succumbing to the stark reality of production physics, declines nominally at 10 percent per year. To replace the lost production and cover the increased demand, new production of 100 million barrels per day must be *activated* (or *reactivated*, if production was shut-in or postponed) over the next decade. Making matters more demanding, the required annual additions become larger and larger with time, from 9 million barrels per day in the first year to 9.5 million in the fifth year and 10.4 million barrels per day in the 10th year.

These figures put into perspective the overstated popular notions in 1999 of excess OPEC capacity, estimated to be no more than 4 million barrels per day.

The investment will be formidable. The average activation index of the major exporting countries is $3,500 per barrel per day of stabilized production. Allowing for a 5 percent annual escalation (in constant dollars), activation costs in the fifth year of the forecast period would exceed $40 billion, increasing to $57 billion in the 10th year. Keeping up with oil demand over the next decade becomes a $425 billion exercise.

With a similar magnitude of investment required to meet the demand for natural gas, total spending will exceed $1 trillion over the next decade.

It Will Be Even More Difficult Later (But Do Not Despair)

Predictions of the future supply of petroleum have typically been far less veracious than predictions of demand. Flawed predictions have caused public bewilderment, distrust, and more importantly, government inaction or poorly conceived reactions (see *Part VII—Yellow*). A constant theme, from the very infancy of the industry at the turn of the 19th century, has been that the world is running out of oil. This worried governments in the 1930s and 1940s and became the focus of a giant public discourse following the energy crisis of 1973.

Part of the problem is the natural and understandable, but often wrong, tendency to predict the future based on the past—what one knows and how things behaved previously. Further aggravating the case for petroleum is the invisibility of the resource, its highly capricious distribution, its actual magnitude when it is discovered, and the recovery factor assigned to it. Petroleum geology, exploration and the physics of production are explained in *Part II—Black*.

Terms such as *oil-in-place* and *reserves*—themselves subdivided into categories such as *proven*, *possible*, *probable*, *recoverable* and *economically recoverable*, and with values that are constantly changing—have added to the confusion, not just for the public, but even for industry workers. In the mix of physical production constraints and modern business practices, these confusing definitions can translate to almost arbitrary book values. Reported book values can be affected by any number of factors, from political to economic to even personal motivations of not risking being proved wrong. At times, book values are contradictory even inside a company, not to mention the reporting problems that exist between governments and countries.

A first attempt can be made to understand the situation by examining cumulative production of oil and gas thus far in the entire history of petroleum.

Cumulative petroleum production to date is 870 billion barrels of oil and 2,300 trillion standard cubic feet of natural gas. This is equivalent to 1,280 billion barrels of oil. (One billion barrels of oil has an energy content that is equivalent to 5.55 trillion standard cubic feet of gas.)

In 1997, Wolfgang Schollnberger of BP Amoco estimated that the world's ultimate recovery—what has already been produced, plus estimated future production (using a recovery factor of 0.35)—will be 5,700 billion barrels of oil equivalent. This figure includes 2,020 billion barrels of "proven reserves," 510 billion barrels of "reserve additions" from existing fields, and 2,110 billion barrels from undiscovered fields.[5] The 5,700 billion barrels of oil equivalent represents 3,300 billion barrels of oil and 13,800 trillion standard cubic feet of gas.

How sustainable are current oil consumption trends? Given today's 75 million barrels per day of oil consumption (27 billion barrels per year) and assuming a 1.8 percent annual increase, worldwide demand would require an additional cumulative recovery of 7,500 billion barrels of oil by the end of the 21st century. This is 2.5 times the current reasonable estimates of ultimate recovery. Such consumption trends are clearly problematic and unattainable.

However, with our scenario of a massive transition to natural gas, projected additional cumulative oil consumption by 2020 is only 300 billion barrels, far below the 650 billion barrels suggested by DOE estimates. More importantly, the daily oil consumption drops from 107 million barrels to 55 million barrels, lower than today's consumption rate of 75 million barrels.

After 2020, assuming that oil is used as a source of materials (as we believe it should be) and that the annual rate of increase in consumption is small, current ultimate recovery estimates should ensure the supply of oil for at least 200 more years. Increasing the recovery factor beyond Schollnberger's 35 percent[5] could extend this period by 100 years.

We predict that the world will not run out of oil for the next three centuries, at least.

The scenario for natural gas is even more optimistic. Using the DOE forecast for annual natural gas consumption (starting at 88 trillion standard cubic feet and increasing by 3.3 percent per year), cumulative natural gas consumption between 2000 and 2020 will be 2,400 trillion standard cubic feet. With our enhanced emphasis on natural gas and the transition to it as a primary fuel, cumulative recovery during the same period will be 2,900 trillion standard cubic feet. All of these figures are well within the current ultimate recovery estimates of 13,800 trillion standard cubic feet.

The total longevity of natural gas is difficult to estimate because it can be produced from coal, as well as produced from reservoirs in its natural state. But we can readily estimate that gas will be the primary fuel and, in turn, the main source of hydrogen (see *Part VIII—New Green*) well into the 22nd century.

Beyond Natural Hydrocarbons

Even with abundant natural gas, the long-term future of energy simply cannot be left to natural hydrocarbons.

A few resources not currently in widespread use can be envisioned playing a big role in the not-too-distant future. We have advocated marked increases in deep offshore petroleum activity and explained the rationale for it in *Part V—Primary Colors*. Another obvious thrust will be for natural gas hydrates, which exist in massive quantities in arctic reservoirs and offshore. Japan is already testing this resource. The use of enhanced oil-recovery techniques, a group of technologies that is proven but currently uneconomical, will allow the re-exploitation of known petroleum resources, substantially increasing reservoir recovery. Coal gasification and coal liquefaction are known processes with proven results.

One or all of these seemingly exotic resources could quickly become conventional, based on shifts in economic conditions, process improvements, and especially, difficult-to-predict step changes in the technologies.

There is little question in our minds that a smooth transition to non-hydrocarbon energy must become a worldwide priority.

We cannot see a substantial future for solar, wind and the so-called "renewable" energies. They may evoke all sorts of emotional responses, but their capacity to provide a sizeable portion of future energy is limited.

We see two potential energy sources for the long-term future: nuclear fission and fusion for stationary energy (i.e., electrical power), and hydrogen for movable energy (i.e., transportation fuel). Hydrogen, propelled into the world economy as a hydrocarbon-derived fuel during the next two centuries, will eventually have to originate from non-hydrocarbons, mainly water.

Such a transition will require the most demanding technology development drive in the history of humankind. Quantum leaps in technological innovation will emerge. In the meantime, hydrocarbons can cushion this transition both by providing very robust interim solutions and by paving the way to the hydrogen-based economy.

From Vision to Implementation

Of course, all visions, far-reaching and rational as they may be, find their implementation more difficult. In an environment where short-term economic considerations reign supreme, the economics of anything new (capital investments) has trouble displacing the economics of the old (maintenance costs). Specific to energy, the question is, "Who will be saddled with the investment required to transform the industry, or to add capacity? How do we pay today for the needs that will arise 10 years from now, especially if it would take that long to prepare for those needs?

The world is not run by government fiat, and experimentations with philosopher-kings have failed. For example, it is not likely that the U.S. government would pay to increase the capacities of facilities needed for importing liquid natural gas into the country. Near-term economic considerations will prevail, and it is certain that growing pains are on the horizon.

Shortages are the most convincing impetus for investment. Consider the current U.S. consumption of natural gas: 22 trillion standard cubic feet per year (18.5 trillion from domestic sources and 3.5 trillion from Canada and Mexico). The average is 60 billion standard cubic feet per day, but average consumption rates can be misleading. Because much of the natural gas is used for space heating, there is a seasonal trend in consumption: a maximum positive peak in winter and a negative peak in the summer, when gas is re-injected and stored for the next winter (Figure 2).

This situation, predictable and logical, has lasted for decades, even though the magnitude of the peaks may have grown over the years. It is about to change.

It is no longer speculation but a fact that effectively all electric power plants planned or under construction in this country will be gas-fired. The growth of gas-fired power generation will be limited only by the capacity of U.S. turbine

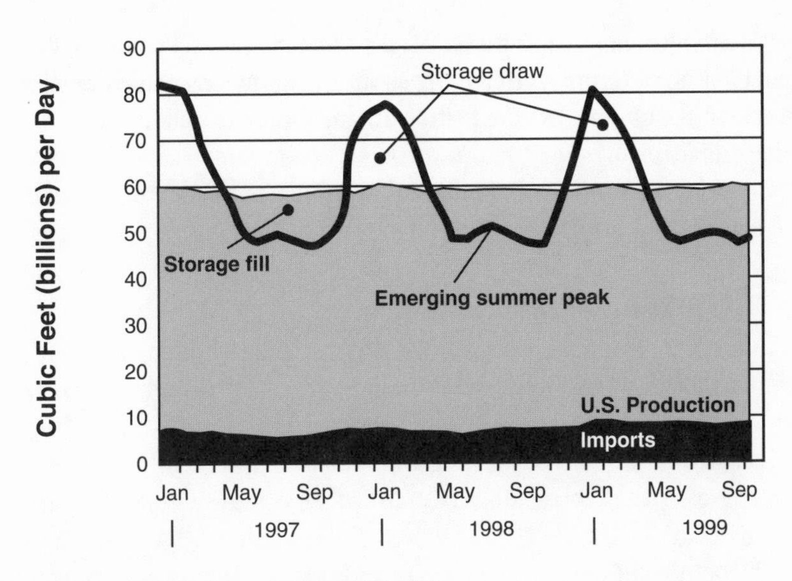

Fig. 2—U.S. gas supply and demand cycle[6]

manufacturers in the foreseeable future. The combined backlog of orders for gas turbines at the start of 2000 suggests that annual U.S. gas consumption will grow by 4.5 trillion standard cubic feet during the next three years.[7] Average daily consumption will climb by 12 billion standard cubic feet per day, a rather dramatic 20 percent increase over current levels.

The result will be profound.

The winter peak will become larger, but the summer negative peak will go away and may even become positive. Air-conditioning demand has its own peak. Natural gas prices will eventually peak twice per year, once in the summer and once in the winter (Figure 2). Depending on seasonal weather variations, regional if not widespread gas shortages will be almost inevitable during the next few years of transition. Gas storage adds both a mitigating and potentially destabilizing element.[8]

Gas shortages will cause power shortages. Brownouts are not just likely; they are certain. And because the U.S. electricity supply is now extensively tied up in so-called "reliability regions," the power grid is inflexible. When shortages hit, at exactly the wrong time, power-hungry consumers (unaware of the consequences of their actions) will tax the system further and exacerbate the problem. A power shortage in New York will affect Los Angeles. On top of that, deregulation, while providing cost benefits to consumers, has created a cut-throat environment among electricity suppliers. Restarting a massively inflexible grid in the absence of cooperation among power companies will not be a simple exercise.

Cassandras we are not, but alerting consumers to the eventualities is both a responsible and fitting gesture.

Transformation of the Business

The information revolution, already transforming the world in ways that were unthinkable only a few years ago, is finding a very fertile ground in the petroleum industry.

First, it allows gifted people in the developing world, from India to China, to become global entrepreneurs, accelerating their country's wealth and, in turn, increasing energy demand.

The petroleum industry, which has traditionally been one of the heaviest users of computing power among all industries, is entering e-trade and e-procurement unambiguously. The global reach that is entailed by the most advanced and newest information technologies fits very well with the eminently global petroleum business.

An industry that has been committed to free trade throughout its history should, perhaps more than any other industry, be expected to eagerly embrace the new mechanisms that break down barriers—be they economic, cultural or political. For example, sanctions imposed on important petroleum nations such as Iran, Iraq and Libya are rapidly becoming ineffective and anachronistic, irrespective of one's attitude toward their original justification.

The new information-led transformation will be all-encompassing, involving trade, services, technology, movement of expertise and on-line contracting. The petroleum industry, so dependent on information, is likely to be the first industry to restructure its management by installing a chief information officer on par with current senior managers such as the chief executive officer and the chief operating officer.

Our Final Words in This Book

We are bullish on the energy industry and incurably optimistic about its future and the future of the world.

If there is one thing that we want to accomplish with this book, it is to educate people on the topic of energy and its impact. Often, it is astounding to discover how energy-illiterate people are; many think that electricity is a form of energy instead of a form of power; others believe that wood-burning accounts for 30 percent of the world energy mix when, in fact, it may constitute less than one-tenth of 1 percent.

The petroleum industry, the current undisputed king of energy, has had and will continue to have an invisible but firm grip on billions of lives.

We are not running out of oil. We are not running out of gas. We do not need to voluntarily reduce our consumption of energy and natural resources.

Energy and its appropriate deployment are the most critical of all wealth-generating activities, and they are the most important modern indicators of the wealth and poverty of nations. Society and energy will merge in an unbreakable bond for the entire future of humankind.

Abundant, cheap energy should always be the goal: *wealth through energy*.

We believe that market forces will naturally generate and marshal new energy resources and prompt the development of new technology, which will lead to appropriate solutions. Governments should stay out of the petroleum business, but they should enable its future by funding critical long-term research on future technologies. In view of the significant time lag between research and implementation, this is an appropriate role for government.

Maintaining a rich energy future will be challenging. It will require the best management and technology that man can master, and it will demand constant attention to cost, but the benefits will be enormous and gratifying. It will also be lots of fun.

To paraphrase a statement by an old friend and colleague, costs must always be controlled, but "the champagne must be vintage."

Acknowledgments

We owe a hearty thanks to Jeanne Perdue and especially Dick Ghiselin for encouraging us to write this book, and for running extended excerpts in their Hart's E&P magazine. The feedback and support have been crucial in bringing this book to fruition, and keeping us on schedule.

We want to thank our friend and colleague, Phil Lewis, specifically for the research he did in support of *Part VII—Yellow,* but also for exhaustive discussions and many of his thoughts that have contributed throughout the book.

With a single line of unedited text, we say thank you to the staff of Metamor Documentation Solutions for their adept and tolerant editing of the entire book. We also thank Douglas Perret Starr, a Pulitzer Prize winner and journalism professor at Texas A&M University who has reviewed several pieces of the manuscript on our behalf.

Finally, we want to thank our wives, Christine Ehlig-Economides and Cheryl Oligney. In addition to being our first audience and most faithful critics, they get credit for keeping us out of trouble.

Bibliographical References

Part I—Green

1. Mackey, S.: *The Saudis, Inside the Desert Kingdom,* Meridian, New York, 1988.
2. "Statistical Review of World Energy," BP Amoco, June 1999, <http://www.bpamoco.com>.
3. "Selected Country Profiles," *The Wall Street Journal Almanac,* Ballantine Books, New York, 1999.
4. "Worldwide Gas Processing," *Oil & Gas Journal,* June 8, 1998, pp. 57-101.
5. "1998 Worldwide Refining Survey," *Oil & Gas Journal,* December 21, 1998, pp. 49-92.
6. "Worldwide Production," *Oil & Gas Journal,* December 28, 1998, pp. 40-68.
7. *The Fortune 500 List,* 1999 <http://www.pathfinder.com/fortune/fortune500/500list.html>.
8. "Crude and Products Prices," *Oil & Gas Journal,* January 25, 1999, p. 55.
9. "Petroleum Engineering Education: Boom & Bust," Department of Petroleum Engineering, Texas A&M University, March 12, 1998, <http://spegcs.org/techtran/boombust.pdf>.
10. "Hot Job Tracks," *U.S. News & World Report Online*, 1999, <http://www.usnews.com/usnews/nycu/work/wohotjob.htm>.
11. *International Energy Annual,* Energy Information Administration, March 1999, <http://www.eia.doe.gov/emeu/iea/wec. html>.
12. "International Energy Outlook 1999," Report No. DOE/EIA-0484(99), Energy Information Administration, Washington, D.C., 1999, <http://www.eia.doe.gov/oiaf/ieo99/>.
13. "Drowning in Oil," *The Economist,* March 6, 1999, pp.19-25.
14. "Shell Reorganizes for Speed and Profit," *Oil & Gas Journal,* December 21, 1998, p. 31.
15. "Shell Chairman Sees Future of 'Structurally Lower' Prices," *Houston Chronicle,* by Agence France-Presse, October 28, 1998, p. 5D.

Part II—Black

1. Pees, S.T. (ed.): *History of the Petroleum Industry Symposium Guidebook: Titusville,* American Association of Petroleum Geologists, 1989.
2. Schlumberger, A.G.: *The Schlumberger Adventure*, Arco Publishing, New York, 1982.
3. Rodengen, J.L.: *The Legend of Halliburton*, Write Stuff, Fort Lauderdale, FL, 1996.
4. Hubbert, M.K.: "Techniques of Prediction as Applied to the Production of Oil and Gas," *Proceedings,* symposium of U.S. Department of Commerce, National Bureau of Standards, Washington, DC, June 18-20, 1980.
5. Petroleum Development Oman L.L.C., Muscat, Oman, 1998.
6. Lewis, P.E.: Oil production and reserve data compiled from various sources, Tulsa, OK, 1999.
7. Brett, J.F. and Feldkamp, L.D.: "The Evidence for and Implications of a Fractal Distribution of Petroleum Reserves," paper SPE 25826, 1993.
8. Oklahoma Corporation Commission, Oil & Gas Conservation Division, Statistical Division, 1999, <http://www.occ.state.ok.us/TEXT_FILES/o&gfiles.htm>.

Part III—Red, White and Blue

1. Chernow, R.: *Titan: The Life of John D. Rockefeller, Sr.,* Random House, New York, 1998.
2. Yergin, D.: *The Prize: The Epic Quest for Oil, Money and Power,* Simon & Schuster, London, 1991.
3. Rockefeller, J.D. (ed.), Porter, Glen (ed.) and Schipper, M.P.: *Papers of John D. Rockefeller, Sr.*, University Press of America, 1991.
4. Ernst, J.W. (ed.): *"Dear Father"/ "Dear Son,"* Fordham University Press in cooperation with Rockefeller Archive Center, New York, 1994.
5. Reich, C.: *The Life of Nelson A. Rockefeller: Worlds to Conquer, 1908-1958*, Doubleday, New York, 1996.
6. Wolf, G.: "I, Monopolist: John D. Rockefeller, America's Original Supercapitalist on Bill Gates and the Microsoft Trial," *Wired,* April 1999, pp. 146-203.
7. "Top 100 U.S. Foundations," The Foundation Center, New York, 1999, <http://fdncenter.org/>.

8. "Exxon, Mobil Chairmen Testify on Deal: Merger is Necessary to Compete on Global Scale, House Panel Told," *Houston Chronicle,* March 12, 1999, p. 2D.
9. "Why Big Oil is Getting a Lot Bigger: Exxon, Mobil, and Rockefeller's Legacy," *U.S. News & World Report,* December 14, 1998, pp. 26-28.
10. "Exxon, Mobil Merger May Require Sale of Assets," *Houston Chronicle,* December 2, 1998, p. 2D.
11. National Petroleum News website, March 1999 <http:// www.npn-net.com/ market98/branded.asp>.
12. "$74B Deal Largest Ever," *USA Today*, December 2, 1998, p. 1B.
13. "Exxon-Mobil Merger Plan Beyond Big: Regulators Worry About Effect of Proposal on Marketing, Refining," *Houston Chronicle*, March 11, 1999, p. 2D.
14. "PIW Ranks The World's Top Oil Companies," *Petroleum Intelligence Weekly—Special Supplement Issue,* December 22, 1997, pp. 1-4.
15. Lowe, J.: *Bill Gates Speaks: Insight from the World's Greatest Entrepreneur,* John Wiley & Sons, New York, 1998.
16. Rohm, W.G.: *The Microsoft File: The Secret Case Against Bill Gates,* Times Books, a division of Random House, Inc., New York, 1998.
17. Stross, R.E.: *The Microsoft Way: The Real Story of How the Company Outsmarts Its Competition,* Addison Wesley Longman, Reading, MA, 1996.

Part IV—Red

1. Giddens, P.: *The Birth of the Oil Industry*, Macmillan, New York, 1938.
2. Yergin, D.: *The Prize: The Epic Quest for Oil, Money and Power,* Simon & Schuster, London, 1991.
3. Gerretson, F.C.: *History of the Royal Dutch*, Brill Publishing Company, Leiden, The Netherlands, 1953.
4. Ferrier, R.W.: *The History of the British Petroleum Company, The Developing Years, 1901-1932*, Cambridge University Press, Cambridge, MA, 1982.
5. Churchill, R.S.: *Winston S. Churchill*, Houghton Mifflin, Boston, MA, 1967.
6. Hart, B.L.: *A History of the World War, 1914-1918*, Faber and Faber, London, 1934.
7. Gilbert, M.: *The Second World War*, Holt, New York, 1989.
8. Hitler, A.: *Mein Kampf*, New York, 1939.

9. Rich, N.: *Hitler's War Aims*, Norton, New York, 1973.
10. Meyer, M.W.: *Japan: A Concise History*, Littlefield, Lanham, MD, 1993.
11. Beasley, W.G.: *The Japanese Experience*, University of California Press, Berkeley, CA, 1999.
12. Freedman, L. and Karsh, E.: *The Gulf Conflict*, Princeton University Press, Princeton, NJ, 1993.

Part V—Primary Colors

1. Economides, M.J. and Oligney, R.E.: "'News' Causes Serious Market Overreactions," *Dallas Business Journal*, October 29-November 4, 1999, p. 63.
2. Wintersteller, W.: "*The Role of Technology in the Upstream Petroleum Industry*," unpublished Ph.D. Dissertation, Mining University Leoben, 1993.
3. "What's the word in the Lab? Collaborate," *Business Week*, June 27, 1994, p. 78.
4. Economides, M.J.: "The State of R&D in the Petroleum Industry," Distinguished Author Series, *Journal of Petroleum Technology*, July 1995, p. 586-588.
5. Foster, R.N.: *Innovation: The Attacker's Advantage*, Summit Books, New York, 1986.
6. Brossard, E: *Petroleum Research and Venezeula's Intevep: The Clash of the Giants*, Pennwell Books/Intevep, 1993.
7. Energy Information Administration website, 1999, <http://www.eia.doe.gov/pub/energy.overview/aer/aer0502.txt>.
8. "An Assessment of the Undiscovered Hydrocarbon Potential of the Nation's Outer Continental Shelf," OCS Report MMS 96-0034, U.S. Department of Interior, Minerals Management Service, 1996.
9. Economides, M.J. and Oligney, R.E.: "Trillion-Dollar Energy Opportunity in Our Grasp," *Houston Chronicle*, July 18, 1999, p. 1C.
10. Bismuth, P. and Kent, J.M.: "Career Management—An International Perspective," invited paper presented at SPE Annual Technical Conference and Exhibition, New Orleans, September 1994.
11. Weidenbaum, M.: "A Key Driver for the U.S. Economy," *Vital Speeches of the Day*, June 1, 1999, p. 506.

Part VI—Colors Of The Rainbow

1. Landes, D.: *The Wealth and Poverty of Nations: Why Some Are So Rich and Some So Poor,* W.W. Norton & Company, New York, 1998.
2. Smith, A.: *The Wealth of Nations,* Alfred A. Knopf, New York, 1776 [Everyman's Library, 1910].
3. Lewis, A.: *The Evolution of the International Economic Order,* Princeton University Press, Princeton, NJ, 1977.
4. Kindleberger, C.: *World Economic Primacy: 1500 to 1990,* Oxford University Press, New York, 1996.
5. Olson, M.: *The Rise and Decline of Nations,* Yale University Press, New Haven, CT, 1982.
6. Granell, E., Garaway, D. and Malpica, C.: *Managing Culture for Success: Challenges and Opportunities in Venezuela,* Ediciones IESA, Caracas, 1997.
7. Sowell, T.: "Race, Culture and Equality," *Forbes*, October 5, 1998, pp. 152-159.
8. Trompenaars, F.: *Riding the Waves of Culture,* The Economist Books, 1993.
9. Nyrop, R. (ed.): *Saudi Arabia: A Country Study,* U.S. Government Printing Office, 1985.
10. Nyrop, R. (ed.): *Persian Gulf States: Country Studies,* U.S. Government Printing Office, 1985.
11. Barwani, M., Marhubi, A., Oligney, R.E. and Economides, M.J.: "*The Role of the Local Petroleum Service Company in Asset Management,*" paper SPE 59445, Yokohama, Japan, April 25-26, 2000.

Part VII—Yellow

1. Rothbard, M.: "War Collectivism in WWI," *A New History of Leviathan*, R. Radosh and M. Rothbard (eds.), E.P. Dutton, New York, 1972.
2. Bradley, R.L., Jr.: *Oil, Gas, & Government*, Rowman & Littlefield Publishers, Landham, MD, 1996.
3. Knowles, R.: *The Greatest Gamblers*, McGraw-Hill, New York, 1959.
4. *National Petroleum News*, March 29, 1933, p. 8.
5. Harold Ickes, Petroleum Administration Board administrator, as quoted in *National Petroleum News*, March 21, 1934.
6. *Oil Week*, July 24, 1933, p. 9.

7. Yergin, D.: *The Prize: The Epic Quest for Oil, Money and Power,* Simon & Schuster, London, 1991.
8. *Oil and Gas Journal*, October 22, 1973, pp. 11-14.
9. *Newsweek*, November 19, 1973, p. 130.
10. Compiled from: Energy Information Administration website, 1999, <http:/ /www.eia.doe.gov/>; and Bradley, R.L., Jr.: *Oil, Gas, & Government*, Rowman & Littlefield Publishers, Landham, MD, 1996, pp. 522-527.
11. "Minus Centi-Millionaire No More," *Forbes*, October 26, 1987, p. 8.
12. "The Rise and Well-Cushioned Fall of Robert Sutton," *Forbes,* August 1, 1983, pp. 34-39.
13. *National Petroleum News*, September 1976, p. 37.
14. *Oil & Gas Journal*, July 25, 1994, p. 60.
15. Tussing, A. and Barlow, C.: "The Rise and Fall of Regulation in the Natural Gas Industry," *Public Utility Fortnightly*, March 4, 1982, p. 15.
16. "U.S. Finds Buyer for Big Synfuels Plant But Won't Recoup Its Initial Investment," *Wall Street Journal*, August 8, 1988, p. 5.
17. "Rogue White Elephant," *Fortune,* June 27, 1983, p. 36.
18. Christensen, M: Former employee at the Great Plains Coal Gasification Project, November 1999, personal communication.
19. "A Federal Reserve for Oil," *Forbes*, March 4, 1991, p. 86.
20. Compiled from: Energy Information Administration website, 1999, <http:/ /www.eia.doe.gov/>; and Bradley, R.L., Jr.: *Oil, Gas, & Government*, Rowman & Littlefield Publishers, Landham, MD, 1996, pp. 1911-1918.
21. Vietor, R.: *Energy Policy in America Since 1945,* Cambridge University Press, New York, 1984.

Part VIII—New Green

1. Bibb, P.: *Ted Turner: It Ain't as Easy as It Looks,* Johnson Books, Boulder, CO, 1993.
2. Gore, A.: *Earth in the Balance: Ecology and the Human Spirit,* Penguin Books, New York, 1993.
3. Lindzen, R.S.: "Global Warming: The Origin and Nature of Alleged Scientific Consensus," *Environmental Backgrounder*, June 18, 1992, pp. 1-12.
4. Coffman, Michael S.: *Saviors of the Earth? The Politics and Religion of the Environmental Movement,* Northfield Publishing, Chicago, 1994.

5. Graedel, T.E. and Crutzen, Paul, J.: *Atmosphere, Climate, and Change,* Scientific American Library, New York, 1995.
6. Mitchell, G.P.: "Temperatures Rising," *Houston Chronicle,* November 23, 1997, p. 1C.
7. Halbouty, M.T. and Westbrook, G.T.: "Before Risking Economy, Prove Earth Is Warming," *Houston Chronicle,* November 23, 1997, p. 1C.
8. Robinson, A. and Robinson, Z.: "Science Has Spoken: Global Warming Is a Myth," *The Wall Street Journal,* Dec. 4, 1997.
9. Singer, S.F.: *Hot Talk Cold Science: Global Warming's Unfinished Debate,* The Independent Institute, Oakland, CA, 1998.
10. Gelbspan, R.: *The Heat Is On: The Climate Crisis, the Cover-Up, the Prescription,* Perseus Books, Reading, MA, 1997.
11. Moore, T.G.: *Climate of Fear: Why We Shouldn't Worry about Global Warming,* Cato Institute, Washington, D.C., 1998.
12. Spencer, R.: "Measuring the Temperature of Earth From Space: Even with Needed Corrections, Data Still Don't Show the Expected Signature of Global Warming," *NASA Space Science News,* August 14, 1998, <http://science.nasa.gov/newhome/headlines/notebook/essd13aug 98_1.htm>.
13. Wentz, F.J. and Schabel, M: "Effects of Orbital Decay on Satellite-Derived Lower-Tropospheric Temperature Trends," *Nature,* August 13, 1998, pp. 661-664.
14. Recer, P.: "Our Climate Could Change Quickly," *Houston Chronicle,* October 29, 1999, p. 24A.
15. "Global Environmental Outlook 2000," United Nations Environment Program, September 15, 1999, <http://www.unep.org/unep/ eia/geo2000/>.
16. Colson, C.: "Punishing the Innocent: The Implant Controversy," *Breakpoint Commentary,* November 5, 1999.
17. Dick Morris as quoted by Richard Reeves, *News and Observer,* Raleigh, NC, December 29, 1998.
18. Huber, P.: "America, the Beautiful Carbon Sink," *Forbes,* April 5, 1999.
19. Leslie, J.: "Dawn of the Hydrogen Age," *Wired,* October 1997, pp. 138-148, 191.
20. "Towards a Hydrogen Economy," *The Shell Report 1999,* Shell Oil Company, 1999, <http://www.shell.com/shellreport/issues/issues4b. html>.
21. Beale, J: CH_4 Corporation, Lawrence, MA, December 1999, personal communication.

22. Swain, E.: "U.S. Became Net LNG Importer in 1997," *Oil & Gas Journal*, January 25, 1999, pp. 78-83.
23. Mathis, N.: "Clinton Creates Federal Council for Work on Biomass Products," *Houston Chronicle*, August 13, 1999, p. 2D.
24. Bernard, B: Bruce A. Bernard Consulting, Inc., Houston, TX, November 1999, personal communication.

Part IX—Purple

1. "Global Environmental Outlook 2000," United Nations Environment Program, September 15, 1999, <http://www.unep.org/unep/ eia/geo2000/>.
2. *International Energy Annual,* Energy Information Administration, March 1999, <http://www.eia.doe.gov/emeu/iea/wec. html>.
3. "International Energy Outlook 1999," Report No. DOE/EIA-0484(99), Energy Information Administration, Washington, D.C., 1999, <http://www.eia.doe.gov/oiaf/ieo99/>.
4. *Energy Outlook 1979-1990,* Exxon Company, U.S.A., Houston, December 1978.
5. Schollnberger, W.: "Projections of the World's Hydrocarbon Resources and Reserve Depletion in the 21st Century," *The Leading Edge,* May 1999, pp. 622-625.
6. *Natural Gas Monthly,* Energy Information Administration, Office of Oil and Gas, U.S. Department of Energy, Washington, DC, October 1999, DOE/EIA-0130(99/10).
7. Bernard, B: Bruce A. Bernard Consulting, Inc., Calculations based on U.S. turbine manufacturing capacity numbers supplied by Bechtel Engineering, Houston, TX, November 1999, personal communication.
8. Moroney, J.R. (ed.): *Advances in the Economics of Energy and Resources,* JAI Press, Stamford, CT, 1999.

Index

A

B

C

D

E

F

G

H

I

J

K

L

M

N

O

P

S

T

U

V

W

Y

Z

About the Authors

Michael Economides

has held chaired professorships at prestigious universities in the United States and Austria. He has written seven textbooks and more than 150 journal papers and articles. He is a technical adviser to national oil companies and Fortune 500 companies. Prof. Economides has advised governments on petroleum policy; provided expert commentary on national television; and written many articles in the popular press on oil, energy, international politics and economics. A native of Cyprus and a naturalized U.S. citizen, Economides earned his Ph.D. in petroleum engineering at Stanford University. He lives in Houston.

Ronald Oligney

is an adjunct professor and director of engineering research development at the University of Houston. Previously, he was principal and founding director of the Global Petroleum Research Institute at Texas A&M University. He was vice president of a New York energy concern and negotiated one of the first joint ventures in the former Soviet republic of Kazakhstan. The company he co-founded, Otek Australia Pty Ltd, is a premier environmental contracting firm in Southeast Asia. Prof. Oligney has written for numerous newspapers and specialized publications, including the Houston Chronicle and the Dallas Business Journal. He lives in Houston with his wife and six children.

Photographs by Michael Zaritski